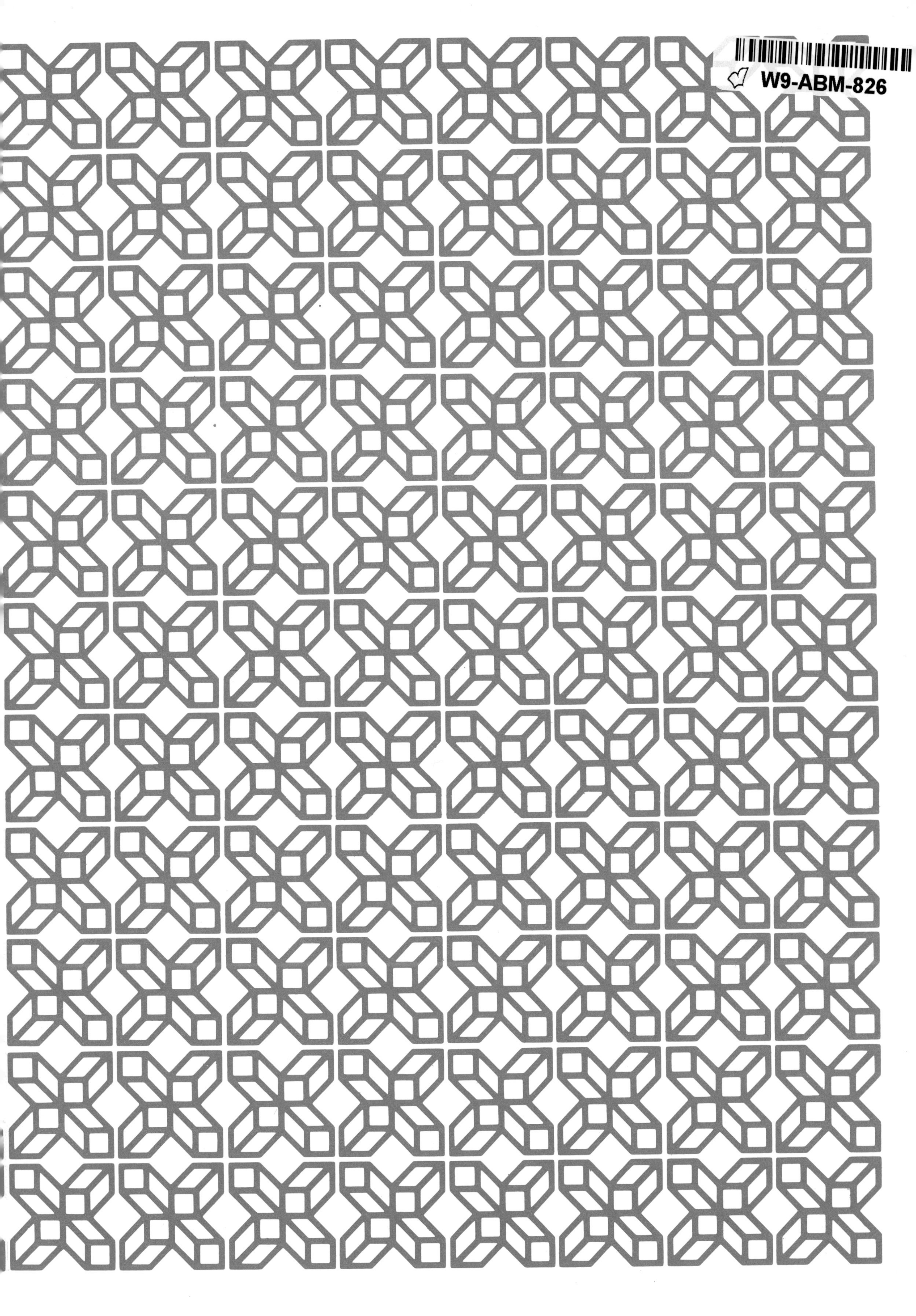

Perspectives in Geography 1

# MODELS OF SPATIAL VARIATION

NORTHERN ILLINOIS UNIVERSITY

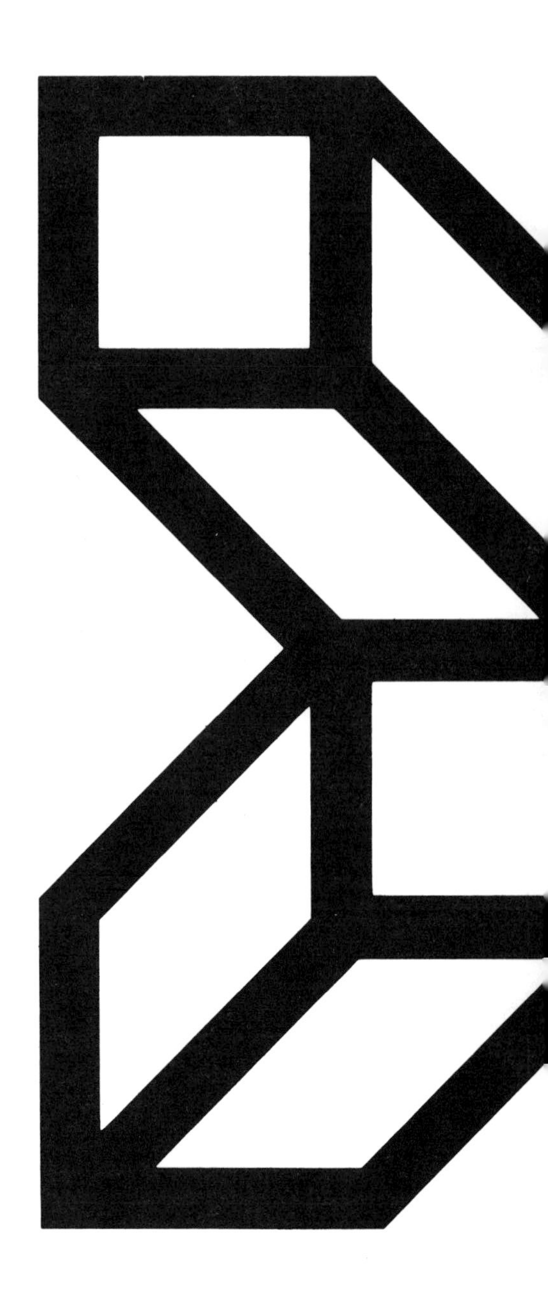

NORTHERN ILLINOIS UNIVERSITY PRESS

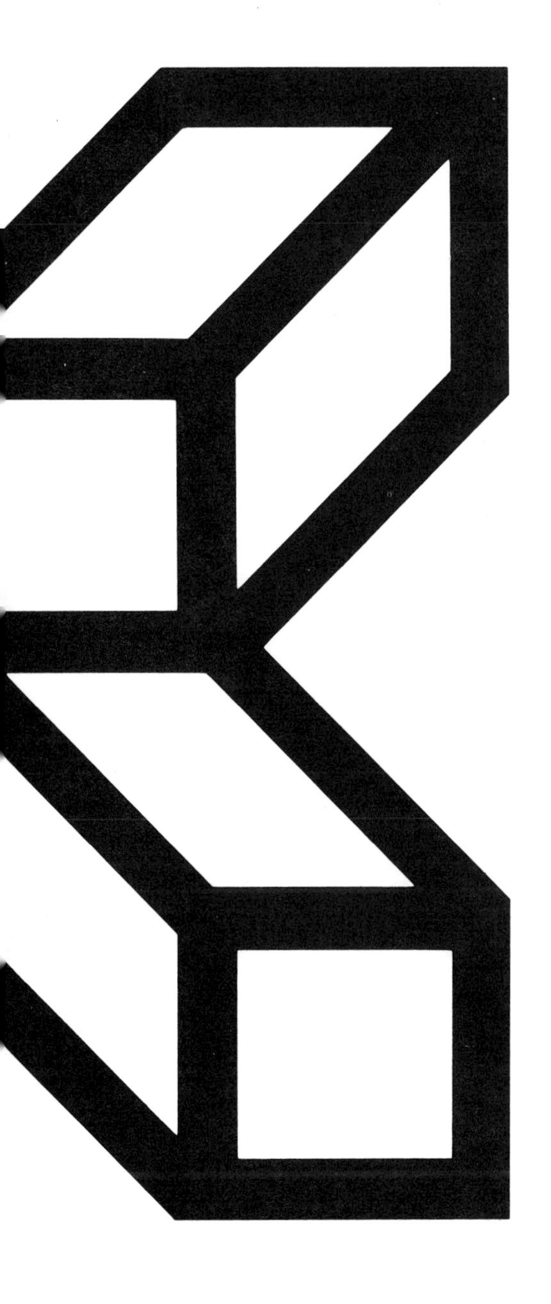

# Perspectives in Geography 1

# MODELS OF SPATIAL VARIATION

EDITED BY

**HAROLD McCONNELL**

AND

**DAVID W. YASEEN**

Harold McConnell is a professor of Geography at Northern Illinois University. David W. Yaseen is an associate professor of Geography at Northern Illinois University.

International Standard Book Number 0–87580–024–6
Library of Congress Catalog Card Number 75–149933
Published by the Northern Illinois University Press,
Dekalb, Illinois 60115
Manufactured in the United States of America

TABLE
OF
CONTENTS

# INTRODUCTION

**Brian J. L. Berry**

The initiation of Northern Illinois University's annual *Perspectives in Geography* with a group of papers dealing with models of spatial variation marks an important departure in the discipline. For the first time, a graduate department has determined that an important service may be made to the field by commissioning a set of papers each year focused on a topic which, though long of traditional concern and interest to geographers, is being transformed in approach and understanding by new modeling strategies.

*Models of Spatial Variation* exemplifies this combination par excellence, for the concern of the book is one of geography's basic concerns—to describe how and explain why phenomena vary spatially. The six papers each contribute one small part to the overall task. Morrill introduces new methods of *pattern analysis,* categorizing the arrangement and measuring the concentration of points in space. Hudson looks at the *diffusion processes* giving rise to spatial patterns. Brown and Albaum model the public *decision-making procedure* lying behind one such diffusion process. Horton and Reynolds examine the *action-space* of individual decision-making. And in the final essay, for a change of pace, I ask of one method of *multivariate pattern-recognition* whether the results produced have substantive utility or are merely artifacts of the analytic method utilized.

The full explanatory cycle implied is a complex one —one that is yet to be achieved in geography, for it involves understanding the decision-making of individuals and institutions within their perceived and actual action-spaces, the accumulation of sets and sequences of decisions into spatial processes, and the manifest outcomes of these processes in patterns of spatial variation. There can be no pretense that the papers in this volume come close to such a cycle; each is but a part of a part. Thus, the contributions made are not those of dramatic new discoveries; rather, they are painstaking increments to understanding that are the hallmark of the bulk of normal scientific activity.

If there is a distinguishing feature of the papers, it is that each facet of the full explanatory cycle is addressed; and this is surely unusual in geography. Certainly it has not been characteristic of quantitative approaches to the field, many of which have gained generality and precision at the cost of myopia and even, some would say, tunnel vision. That breadth as well as depth is now evident speaks for the corrective effects of time and greater maturity of more mathematically inclined geographers. That a strong self-critical flavor emerges, too, is another sign that quantitative geography is coming of age. *Perspectives in Geography* offers an important new avenue whereby the discipline may benefit therefrom.

# 1

# A TEST OF DIRECTIONAL BIAS IN RESIDENTIAL MOBILITY

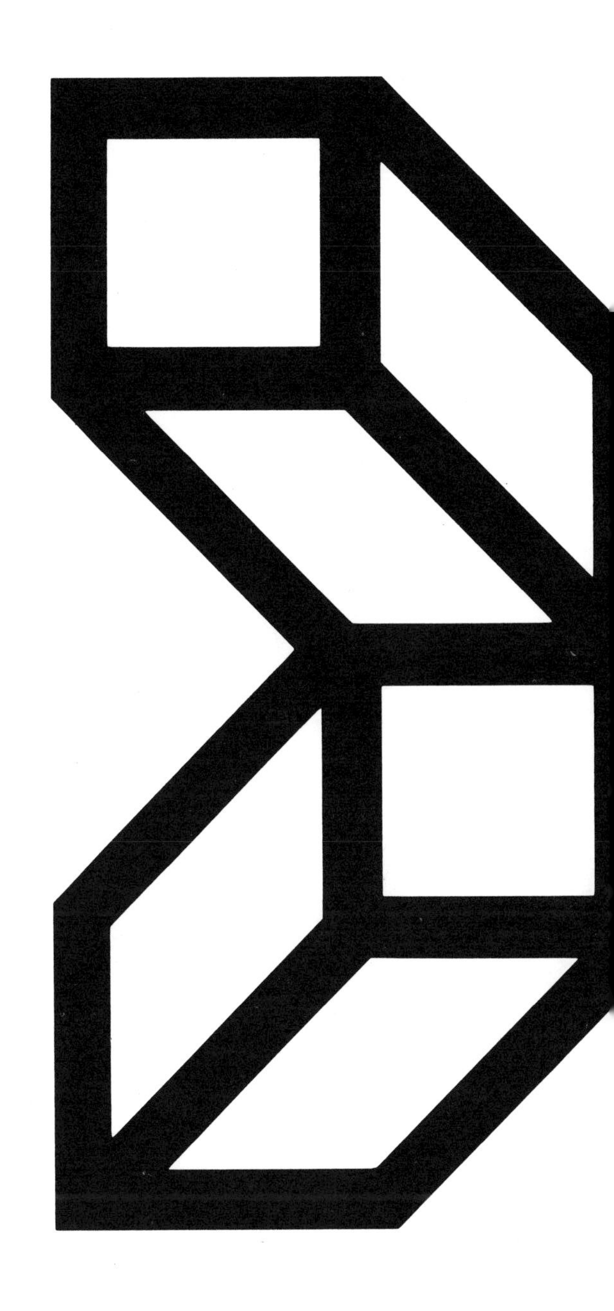

**W. A. V. Clark**
University of California, Los Angeles

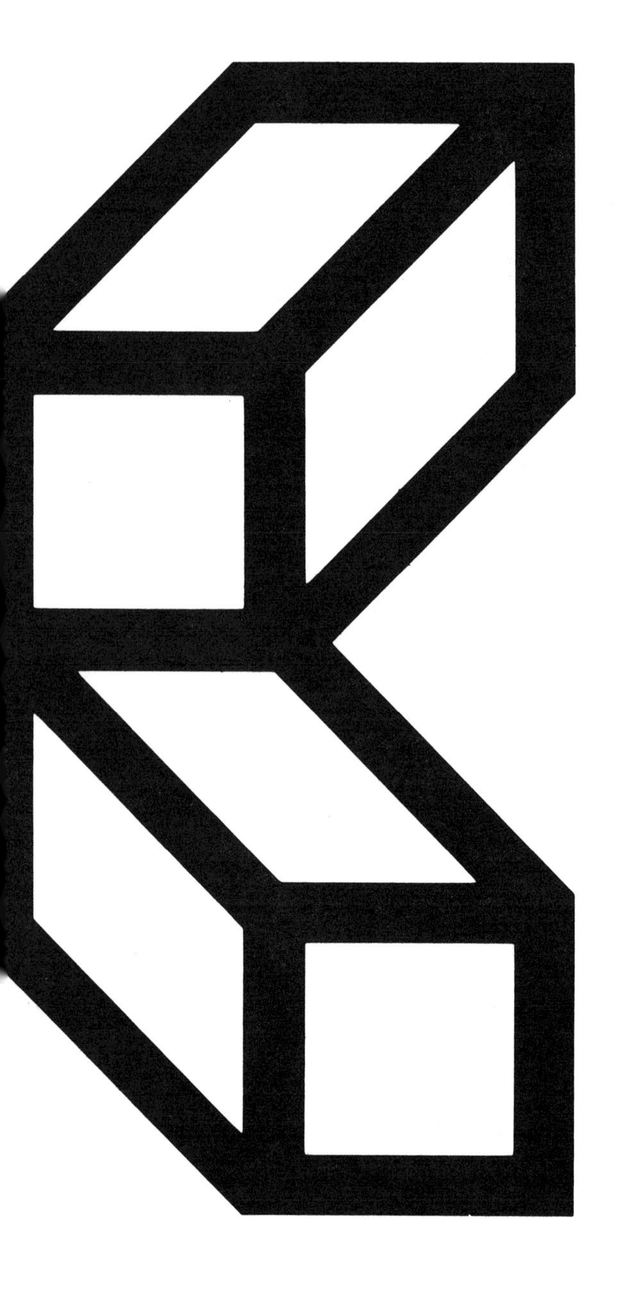

# A Test of Directional Bias in Residential Mobility

## ABSTRACT

An investigation of residential mobility is undertaken within the framework suggested by recent work on spatial behavior and spatial structure. In particular, the directional component of intraurban mobility is examined for the extent to which directions of mobility are random or biased in a fashion which would suggest the influence of the underlying urban structure. The structural models of the city are reviewed and their possible influences on mobility discussed. Two models of residential mobility are suggested, a random or unbiased model and a sector or biased model. Both models are tested for some 5,000 household moves in a New Zealand city of approximately 250,000 people. The data set consisted of the origins and destinations in street addresses and geographical coordinates for each mover. To facilitate the analysis, the city is divided into thirty-six regions. For each region the directions and distances of the moves are calculated and grouped into 30° and 10° sectors. The generated results from the model of random moves are tested against the observed moves using a Kolmogorov-Smirnov test. For those regions for which the model of random moves is an inadequate explanation, the sector model is examined. The results of the tests, although not conclusive, suggest that the central-area moves are random in direction and the moves from the outer areas of the city are biased in a sectoral fashion. It is suggested that the explanation of this dichotomy lies in the distribution of available housing opportunities within the city.[1]



1. I would like to thank Philip Lankford, and Martin Cadwallader of the Department of Geography, U.C.L.A., for comments on an earlier draft of this paper, and Karl Auerbach for technical assistance with the computer programs.

A recent paper has initiated a concern with the directional component of residential mobility, and emphasized its importance by associating direction of movement and urban structure (Adams 1969). This approach concerning itself with structure and behavior reflects a trend in locational analyses that can be identified in recent investigations of human spatial behavior (Rushton 1969).

Many of the models which have been developed to explain the distributions of activities within cities can be described as structural models. For instance, the investigation of social areas within the city is essentially concerned with the spatial expression of types of populations, their covariation with one another, and other physical properties of the urban space. It is reasonable to suggest, however, that a concern with structure alone may be insufficient to understand the total urban or spatial system. In many situations it is necessary to turn to an analysis of behavior *within* a given structure in order to further understand the system. While spatial behavior is often viewed as an observable activity and particularly as the movement of phenomena between places, Rushton (1970) has set up a more useful definition, distinguishing between spatial behavior as the rules by which alternative locations are evaluated and choices made, and behavior in space as the actual spatial choices. Golledge (1970) has added a useful criterion for linking and understanding spatial structures, spatial behavior, and the behavior of structures (or changes in spatial structure). He suggests that the "structural changes over time are the physical manifestations of human behavior . . . the results of individual and group decision making" (Golledge 1970, p. 7). The discussion of structure and behavior is a particularly important one for understanding residential mobility. There is, of course, a structure—the city or subparts of the city—within which observable mobility occurs. There is also a changing urban structure as a response to the temporal spatial movements of the population. As Rossi (1955, p. 1) initially pointed out, it is the residence shifts of urban households which produce a large part of the change and flux of urban population structures and residential areas. While these comments go some way toward linking behavior and structure and, even though the concern in this paper is primarily with structure and behavior within the structure, it is useful to make some additional comments on the way in which behavior is determined.

Research conducted initially by Lynch (1960) and more recently by Stea (1969) has suggested that the images people hold of the city, or subparts of the city, influence their behavior within the city. The individual's information field, or his mental map, becomes of fundamental importance in understanding the behavior of individuals. We might suggest the set of relationships in figure 1. From the figure it can be argued that, while

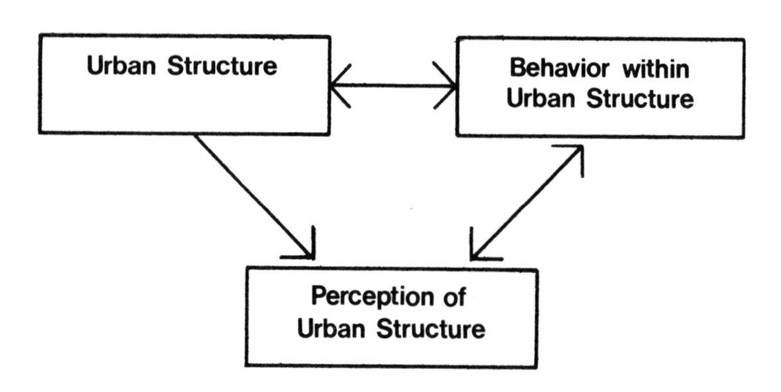


Figure 1. Interrelationships of structure and behavior.

structure and behavior can be associated, it may be through the function of the mental map, image, or perception of the structure that knowledge of the behavioral processes influencing the structure and behavior within the structure can be derived. Behavior and structure are thus linked; or, in a more speculative sense, behavior is converted into structure and structure modifies behavior. This point is also developed by Moore and Brown (1969) who suggest that the analysis of person-to-person contacts and the resulting contact field may be regarded as the first step toward the identification of spatial regularities in information flows. It should be noted, however, that contact fields are not necessarily synonymous with information fields.

At the present time there is only limited information on the perception of urban structures, although there have been preliminary investigations by Lynch (1960), Gould (1966), Moore and Brown (1969), Orleans (1968), and Stea (1969), to mention only selected examples. The attempts to build mental maps and information fields have been hampered by a lack of data collected specifically on the way in which an individual interacts with his surrounding environment. Two approaches to the individual information field and associated mental map have been hypothesized. One model suggests a circular information field declining isotropically from the individual's residential location (Marble and Nystuen 1963; Morrill and Pitts 1967; and Clark 1969). The other suggests a wedge-shaped, elliptical, or biased information field (Adams 1969; Moore and Brown 1969; and Simmons 1968). However, while the distribution of contacts is "not well represented by distance decay functions which are symmetrical around the origin," the evidence suggests that the contact fields are more regular than biased (Moore and Brown 1969). Respondents are differentiated between inner and outer residential locations and, as expected, the inner group of residents seem to have a more even network of contacts than the residents with outer locations. Overall, the form of the contact field might be described as an ellipse with the location of the residence or respondent at one limit of the ellipse. The fact that the set of acquaintances very close to the respondent was excluded may bias the results, although the impact of including these contacts is unclear.

In this analysis the attempt will not be to study the contact field as such, but rather to concentrate on a direct analysis of the urban structure and behavior within the structure. There is insufficient evidence to derive operational information fields at the present time; hence, any evidence on the information field will be indirect only. If, however, the test of the model suggests that the behavior within the structure is random, there is evidence that there is no bias in the information field.

## URBAN STRUCTURE

The structural approaches to the city which have had an explicit spatial component are the Burgess concentric-ring hypothesis and the sector concept of Hoyt. More recent investigations using factor analytic methods have suggested that these two separate approaches can be combined into a composite model (Berry 1965; Murdie 1969; Rees 1969). The factorial investigations have found that socioeconomic characteristics of households vary by sector, and that family structure, family size, and the age structure of the population, on the other hand, vary with distance from the city center or, in other words, roughly concentrically. Localized segregated ethnic groups make up a third and distorting pattern. "If the concentric and axial schemes are overlaid on any city, the resulting cells will contain neighborhoods remarkably uniform in their social and economic characteristics. Around any concentric band communities will vary in their income and other characteristics but will have much the same density, ownership, and family patterns. Along each axis communities will have relatively uniform economic characteristics, and each axis will vary outwards in the same way according to family structure" (Berry 1965, p. 116). The integrated spatial model as drawn by Murdie and Rees is illustrated and suggests the term *wedge shape* (figure 2).

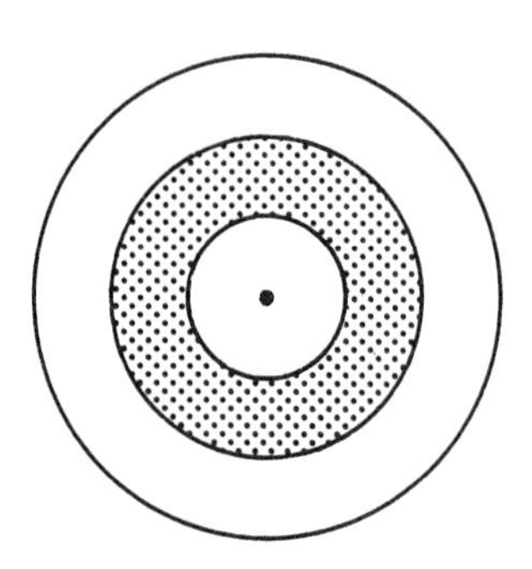

BURGESS-Concentric Ring

HOYT-Sector

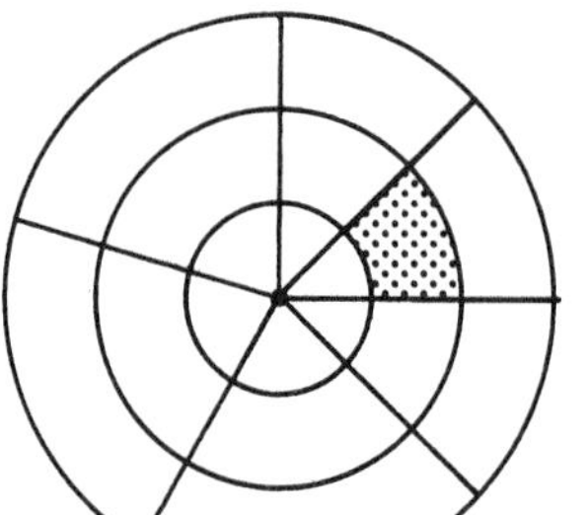

Factorial

Figure 2. Structural models of the city.

Although the factorial model has been given some support by the empirical work of Murdie (1969) and Rees (1970), the maps of factor scores in Rees (1970, p, 359–61) do not wholly and clearly support the models of overlapping rings and sectors. In addition, in a discussion of the socioeconomic factor, Rees notes, "a complex exists of sectors, semisectors, sectors that are almost rings . . . " While this comment does not invalidate the suggestions of an additive model, it raises questions as to whether we can expect clear associations between movements and structural economic sectors.

One view of the process whereby the urban spatial structure has been created by spatial behavior is suggested as that of the individual inhabitant evaluating residential opportunities and personal constraints. The process of choosing a residence is constrained in three ways: the cost of the housing unit, the location of the unit with respect to facilities in the city, and the size of the unit. These constraints, of course, are related to income, family life style, and stage in the family life cycle. Rees (1970) argues that income is the most important determinant in the housing choice. Thus, the move of the individual is constrained by his income and by the areas available to him. As a consequence, the structure of the city (if we accept a definite sectoral pattern of economic characteristics) may reinforce the suggested wedge-shaped image derived from the in-out commuting structure of cities which have a strong central business district (Adams 1969).

The procedure of residential choice is more complicated than that suggested by the simple analysis of constraints. For a detailed discussion of the interrelationships of the awareness space, the activity space, and the search space, see Brown and Moore (1968). Rees (1970) in his discussion of the residential choice procedure suggests a similar model in which the individual occupies a social space and matches this position with a dwelling in an analogous housing space and community space. But since the interest here is to analyze the relationships of the observed decision and the urban structure, rather than to investigate the procedure of residential selection *per se,* we need not develop in detail the decision-making procedure.

In another attempt to link structure and behavior, Adams suggests a biased mental map or image of the city, based on radial travel patterns of urban residents to and from the downtown area. In addition, he suggests a sectoral pattern of movement, reflecting Hoyt's assumptions of moving sectorally to maintain certain socioeconomic characteristics. Although Adams suggests that movement will be inward and outward for *radically different* housing, Berry and Rees suggest a lateral movement for different housing. The contradiction can be explained in terms of the emphasis on the concentric addition of housing units by Adams. Adams and Berry emphasize different elements of the structure of the city. Adams emphasizes the wedge-shaped image of the urban resident drawn from a radial commuting and traveling pattern and a resulting inward-outward mobility pattern. Rees and Berry emphasize income as a major determinant in intraurban mobility and this suggests a sectoral pattern of movement or, in other words, a directionally biased movement.

Although in analyses of the reasons given for residential relocation, life-cycle forces account for 50 per cent of the reasons, and economic forces account for approximately 25 per cent of the reasons, for moving, there is other evidence to support a claim of economic forces as being the most important forces in the residential relocation process. For example, of 67 persons classified as low-income earners, 80 per cent moved to areas classified as low-income areas. Further, these same low-income areas attracted only 15 per cent of the remainder of the sample. Thus, in an analysis of the types of areas within the city and the flows of population into these areas, those differentiated according to economic characteristics can be considered to be sustained by the process of residential mobility. On the other hand, life-cycle factors, which play such an important role in stimulating movement, play a much less important role in determining the final destination of an intraurban move for most households. The stage in the life cycle appears to yield only a relatively small amount of information on the household's final destination (Clark and Stanman 1970).

## MODELS OF MOVEMENT

From the discussion of urban structure and spatial models of urban structure we might expect either the sectoral pattern of the city (economic forces), or the overlapping of the concentric ring and sector models and the resulting cells, segments, or wedges, to dominate movement. In the one case we would expect a biased pattern of movement and in the other, a random pattern. These segments, or cells, are in fact relatively homogeneous neighborhoods and residential mobility will occur within these segments. In addition, it seems likely that any group of families will have varying emphases on the importance of life-cycle and socioeconomic influences and constraints. As a corollary, we might expect that potentially mobile families will investigate several areas rather than look in one particular direction. Thus, an initial research hypothesis in the investigation was that a pattern of random movements was more likely than a biased pattern.

Two models are suggested from the above discussion. One model describes directions of movements within the city as a random vector model. The other is a sector model. We designate a direction $w_\iota$ from $x_\iota$ $y_\iota$, the original location of a mover, and $p_\iota$ represents a distance from $x_\iota$ $y_\iota$.

(1) For any mover
   a. select $w_\iota$ ($o \leq w_\iota \leq 360$) by a random procedure
   b. select $p_\iota$ in accordance with $p = ab^n$, the Pareto formula

(2) For any mover (with $p = 0.5$)
   a. select $w_\iota$ ($o \leq w_\iota \leq 60$)
   b. select $p_\iota$ in accordance with $p = ab^n$, the Pareto formula

In the present investigation, tests of part a of each model is undertaken. It is apparent that Model 2 is an alternative to Model 1 or vice versa. It is also worth noting that the choice of an angle of less than 60 degrees to represent a sectoral bias is arbitrary, but the literature has never specified the exact nature of a sector (Rodwin 1950).

## TESTS OF THE MODELS

The data used in the tests of the models of residential mobility consists of 5,040 records, each record having the name of a mover, the origin of his move in street address and geographical coordinates, and the destination of his move in street address and geographical coordinates. The data set is for Christchurch, New Zealand, a city of approximately 250,000 persons in the year 1965. The greater part of the city has developed on a basically flat coastal site, with some extensions of the residential area onto the lower portions of an extinct volcanic mass. Thus, apart from the ocean and to a lesser extent the volcanic mass, there have been few barriers to the outward development of the urban area. The street addresses were first plotted on a detailed street and lot map of the city and then converted to coordinate locations with the use of an electronic digitizer. This machine linked to an IBM key punch produces a four digit x and a four digit y coordinate for each origin and destination address.

In order to test the model the distance and direction for each mover was calculated. Simple Euclidean distance was computed

$$(3) \quad d = \sqrt{(x_2 - x_1)^2 + (y_2 - y_1)^2}$$

and direction was computed as:

$$(4) \quad \theta\,(\text{degrees}) = \arcsin y_2 - y_1/d$$

The 90 degree quadrant in which the move fell was calculated by comparing the differences in signs between the x and y coordinates. That is, the angle was computed with the formula giving an angle between 0° and 90° and the actual quadrant 0–90°, 90–180°, 180–270°, and 270–360° was chosen from an analysis of the positive and negative signs of the x and y coordinates.[2] Zero was assigned as due east, and 90° due north. However, absolute directions are less important than comparative directions in testing for directional bias. A grid of cells approximately 1.9 x 1.9 miles was randomly located over the city (figure 3).[3] Within each of the regions, the directions taken by movers were assumed to originate from the center of the region and the number of movers falling in any particular sector (10°, 20°, 30°) can be calculated. As a final step, the proportions of moves (angles of moves) can be compared with the proportions falling in each sector according to the models.

As a preliminary stage in the analysis of the directions of residential movement, angles of movement were grouped into 10° and 30° sectors yielding thirty-six and twelve sectors respectively for each region. A tabulation of moves for each of the 360° was also carried out. The directions of movement are illustrated for the 30° sectors in figure 4, and plotted as a table for the 10° sectors in table 1.[4]

The plotted directions of movement by sector for the thirty-six regions can be analyzed descriptively. There is an easily identifiable zone made up of six regions where all twelve sectors are represented with large numbers of movements. Regions 15, 16, 17 and 21, 22, 23 can be observed to have "opposite" sectors with approximately

2. Program descriptions and program decks for the analysis of the directions of movement are available from the author.

3. The cells are square for the major portion of the urban area, but the northern and eastern row of regions were enlarged to oblong, rather than square, shapes in order to include all movers within the urban area and still keep the number of regions to thirty-six or fewer. It is not expected that the oblong regions have biased the results.

4. For one region (16) there is a small difference in the total number of observed moves for the 30° and 10° sectors. For the 30° sectors, 1,000 moves are shown, and for 10° sectors, 1,016 moves are tabulated. The difference relates to the storage capacity of the programs as they were originally written. The tests are unaffected.

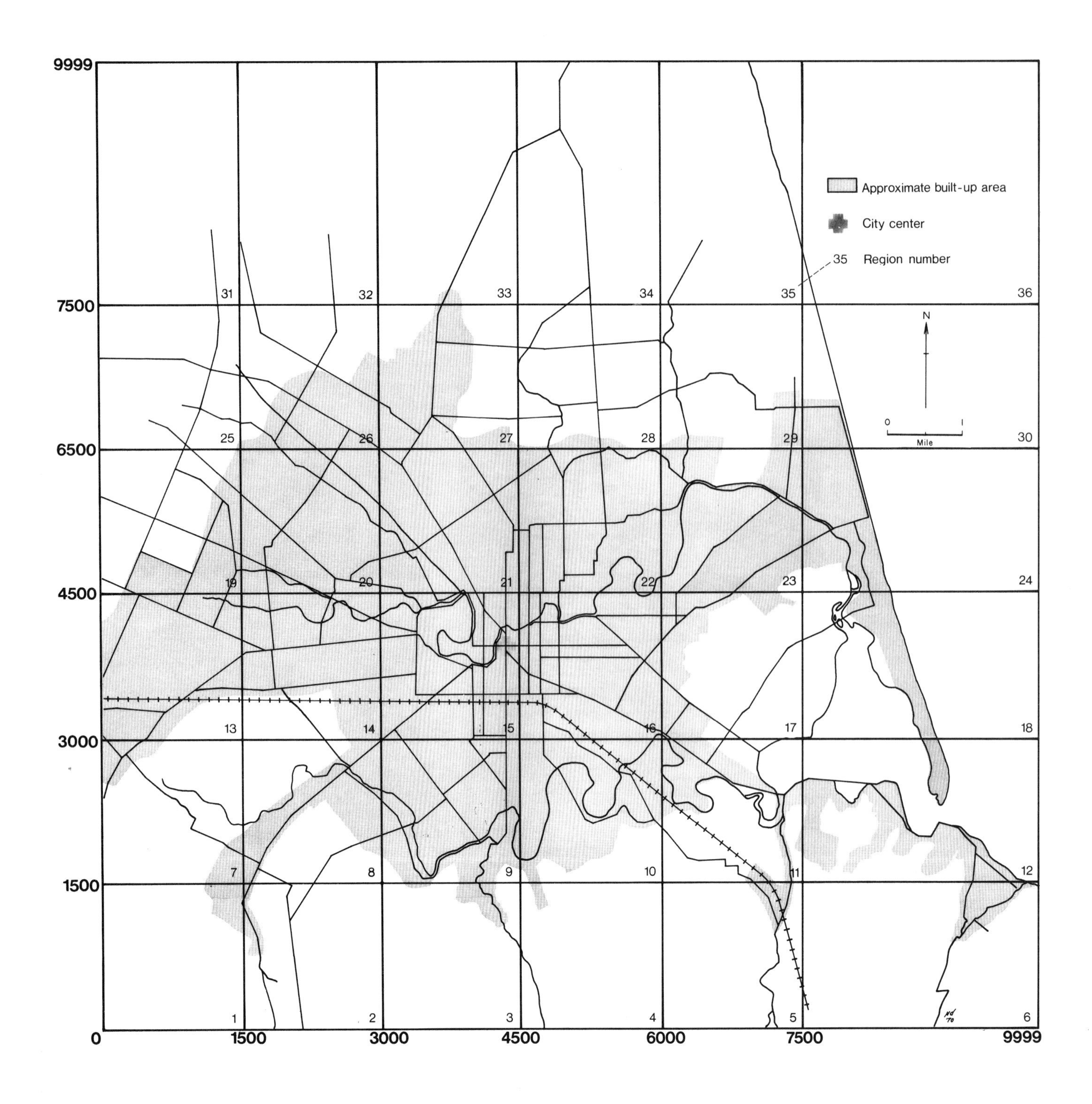


Figure 3. Christchurch Urban Area. Cells are approximately 1.9 × 1.9 miles and are arbitrarily located. Circled figures refer to cell identification numbers used in the text.

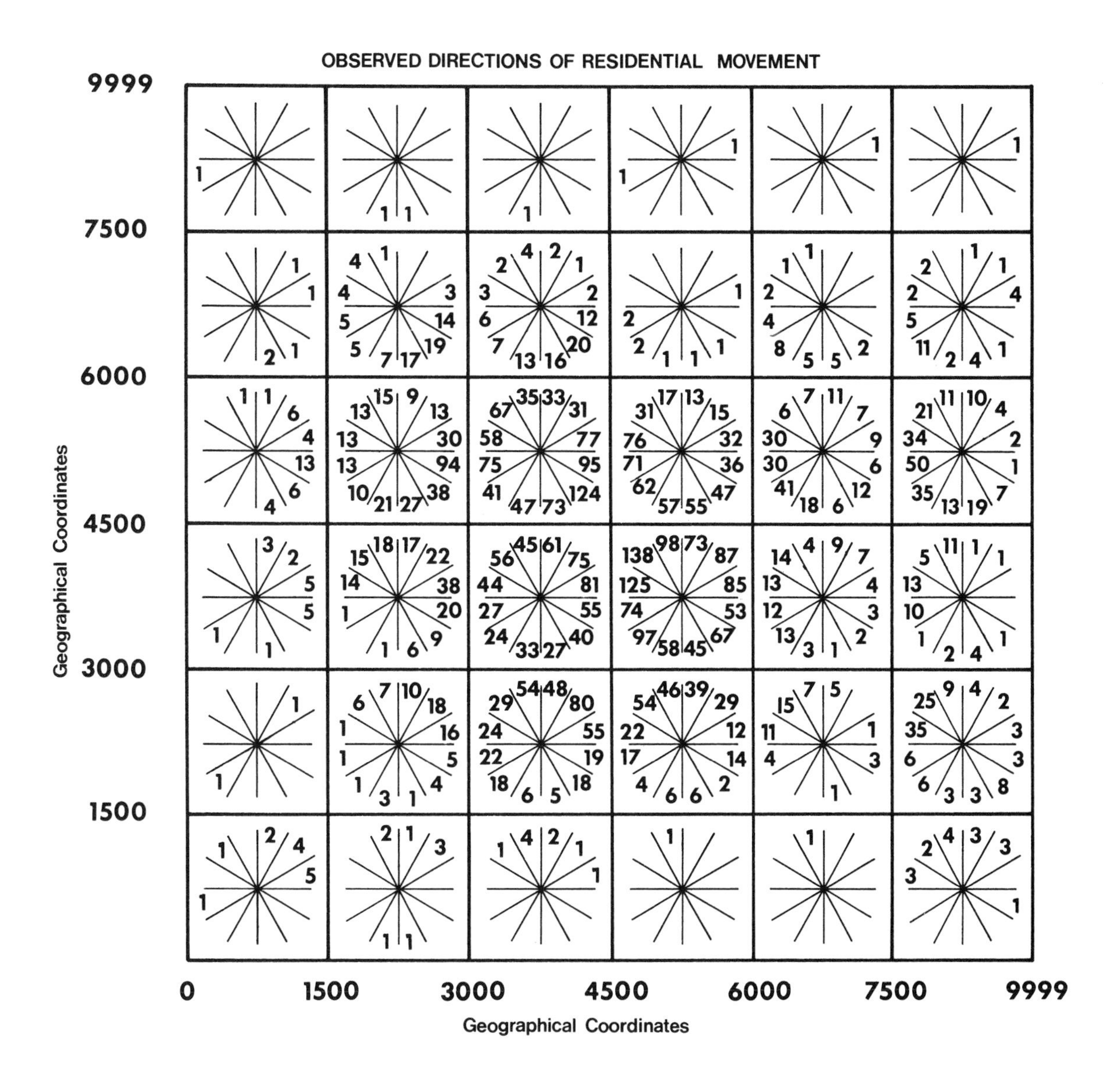


Figure 4. Observed directions of residential movement. The number in each sector indicates the number of households who moved within that sector.

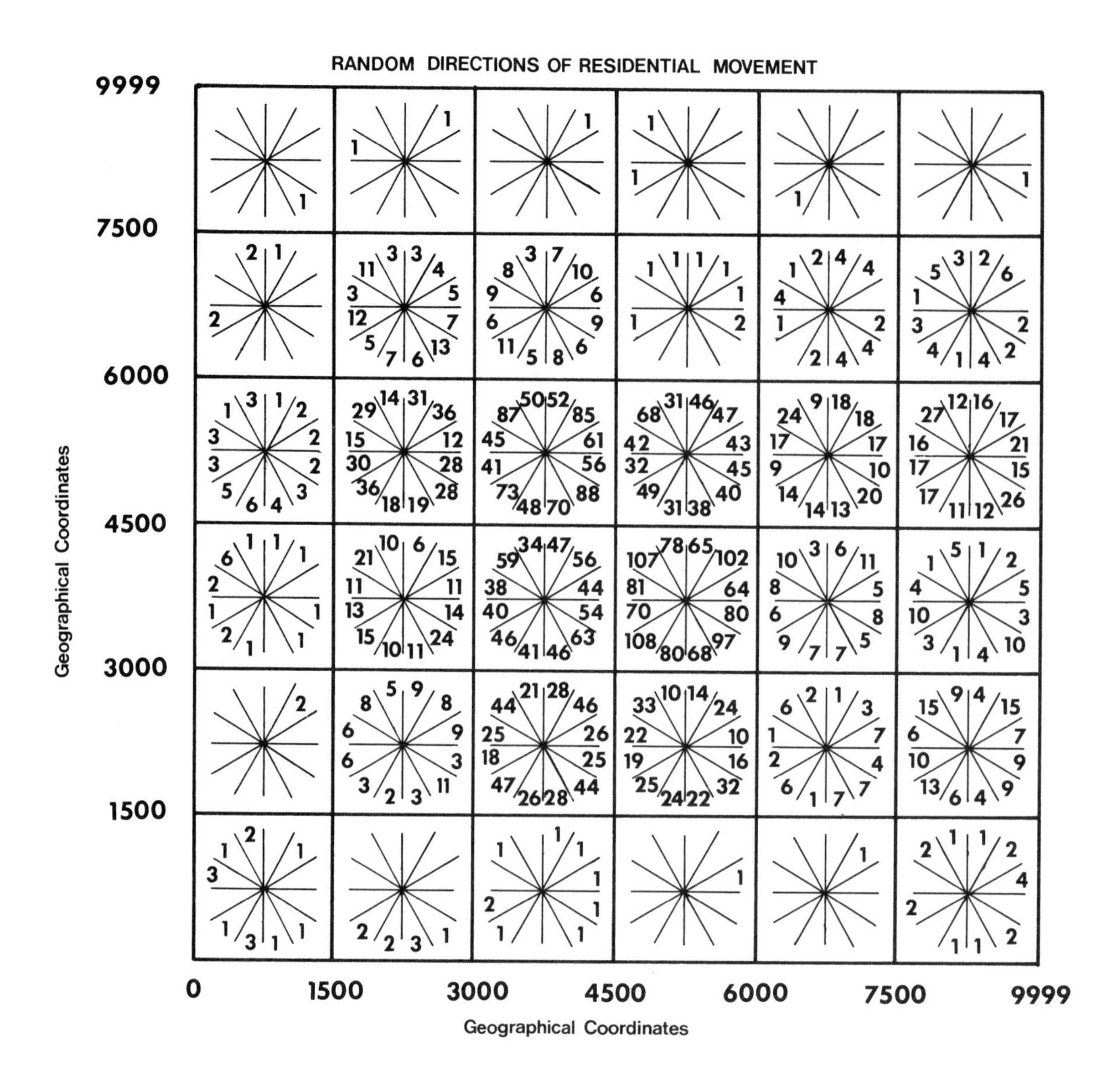


Figure 5. Random directions of residential movement. Numbers generated from Model 2.

the same number of moves. For example, in region 16, the sector between 30° and 60° (when 0° is taken as due east and counting in a counterclockwise direction) has eighty-seven moves, and the sector between 210° and 240° has ninety-seven moves. This central zone of the city is surrounded by regions which can be identified as having a biased pattern of moves; that is, the sectors on one side of the region have a larger number of movements than the sectors on the opposite side. For example, in region 8, the sector between 30° and 60° with eighteen moves is "opposed" by a sector with only one move. The regions surrounding the central zone can be divided into four groups, northwest, northeast, southwest, and southeast. Within each of these groups a major directional component can be identified. The northwest group of regions has moves directed to the southeast, the northeast group of regions has moves directed to the southwest, and the southwest and southeast groups of regions have moves directed to the northeast and northwest, respectively. Thus, of the twenty-seven regions with five or more moves, the moves within the six most central regions are not identifiable as directionally biased, and the moves in the outer regions have a directionally biased component. Similar conclusions can be reached with an analysis of the thirty-six 10° sectors. Regions 15, 16, 17 and 21, 22, 23 have an even spread of the angles of movement throughout the sectors, while regions 9 and 10, to take two examples, have a more clustered structure of angles of movement.

In order to test the models of direction of residential mobility, the angles of movement grouped by the 10° and 30° sectors were compared with angles of movement generated by the model of random movements.[5] The results of the model are illustrated for 30° sectors in figure 5, and for 10° sectors in table 2. To test the extent of correspondence between the angles generated by the model and the actual angles of movement, Kolomogorov-Smirnov one sample statistics were calculated (table 3). The values given in the table are the maximum differences between the observed and generated angles of movement for Model 1 of each sector. The results of the tests generally follow the descriptive outline of differences between the actual number of moves per sector and the number of random moves per sector. There is no significant difference between the actual number of moves of each angle and the number generated by the model of random directions of movement for the tests of either the 10° or the 30° sectors of regions 12, 15, 16, and 17, and those of 21, 22, 23, 24, and 30. Some regions show up as significantly different from random when the thirty-six sectors are considered but not when twelve sectors are analyzed. However, in almost all cases the maximum differences for the regions which have conflicting significance results are larger than for the regions

5. Since the model-generated results are only one set of several possible generations, it seemed that significance tests were needed, even though the total population of moves, rather than a sample, is employed in the analysis.

Table 1. *Observed Directions of Residential Movement.* Each entry in the table represents the number of households with angles of movement falling in that sector.

| Region | Sector (10 degree) | | | | | | | | | | | | | | | | | | | | | | | | | | | | | | | | | | |
|---|---|---|---|---|---|---|---|---|---|---|---|---|---|---|---|---|---|---|---|---|---|---|---|---|---|---|---|---|---|---|---|---|---|---|---|
| | 1 | 2 | 3 | 4 | 5 | 6 | 7 | 8 | 9 | 10 | 11 | 12 | 13 | 14 | 15 | 16 | 17 | 18 | 19 | 20 | 21 | 22 | 23 | 24 | 25 | 26 | 27 | 28 | 29 | 30 | 31 | 32 | 33 | 34 | 35 | 36 |
| 1 | 1 | 1 | 3 | 3 | 1 | | | 2 | | | | | | 1 | | | | | 1 | | | | | | | | | | | | | | | | | |
| 2 | | | | 1 | | 2 | | 1 | | 1 | | 1 | | | | | | | | | | | | | 1 | | | 1 | | | | | | | | |
| 3 | | | 1 | | 1 | | | 1 | 1 | 1 | 1 | 2 | 1 | | | | | | | | | | | | | | | | | | | | | | | |
| 4 | | | | | | | | | | | 1 | | | | | | | | | | | | | | | | | | | | | | | | | |
| 5 | | | | | | | | | | | 1 | | | | | | | | | | | | | | | | | | | | | | | | | |
| 6 | | | | | 1 | 2 | 1 | 2 | | 1 | 2 | 1 | 1 | | 1 | 1 | 1 | 1 | | | | | | | | | | | | | | 1 | | | | |
| 7 | | | | | 1 | | | | | | | | | | | | | | | | | | | 1 | | | | | | | | | | | | |
| 8 | 3 | 9 | 4 | 11 | 3 | 4 | 3 | 4 | 3 | 3 | 2 | 2 | 3 | 2 | 1 | | 1 | | | 1 | | 1 | | | | 2 | 1 | | | 1 | 2 | 1 | 1 | | 2 | 3 |
| 9 | 20 | 18 | 17 | 26 | 27 | 27 | 22 | 11 | 15 | 17 | 21 | 16 | 18 | 8 | 3 | 12 | 5 | 7 | 5 | 13 | 4 | 7 | 6 | 5 | 1 | 4 | 1 | 4 | 1 | | 2 | 8 | 8 | 1 | 11 | 7 |
| 10 | 5 | 3 | 4 | 9 | 8 | 12 | 11 | 9 | 19 | 14 | 19 | 13 | 25 | 15 | 14 | 11 | 5 | 6 | 5 | 7 | 5 | 1 | 1 | 2 | 1 | 1 | 4 | 2 | 1 | 3 | | 1 | 1 | 2 | 5 | 7 |
| 11 | 1 | | | | | | 1 | 2 | 2 | 1 | | 6 | 7 | 2 | 6 | 3 | 3 | 5 | 3 | | 1 | | | | | | | | | 1 | | | | | 1 | 2 |
| 12 | | 2 | 1 | 1 | 1 | | 1 | 2 | 1 | | 5 | 4 | 6 | 4 | 15 | 17 | 11 | 7 | 2 | 2 | 2 | 2 | 1 | 3 | 2 | 1 | | | 1 | 2 | 1 | 2 | 5 | 1 | | 2 |
| 13 | 1 | 2 | 2 | 1 | 1 | | 3 | | | | | | | | | | | | | | | | | 1 | | | | | | 1 | | | | | 2 | 3 |
| 14 | 16 | 13 | 9 | 9 | 10 | 3 | 5 | 4 | 8 | 5 | 8 | 5 | 7 | 4 | 4 | 7 | 4 | 3 | | 1 | | | | | | | 1 | 3 | 1 | 2 | 2 | 4 | 3 | 5 | 5 | 10 |
| 15 | 25 | 15 | 41 | 28 | 23 | 24 | 21 | 25 | 15 | 13 | 16 | 16 | 24 | 18 | 14 | 17 | 11 | 16 | 12 | 6 | 9 | 6 | 12 | 6 | 15 | 8 | 10 | 12 | 7 | 8 | 9 | 12 | 19 | 14 | 22 | 19 |
| 16 | 23 | 36 | 28 | 39 | 27 | 21 | 24 | 25 | 24 | 29 | 28 | 42 | 44 | 39 | 57 | 49 | 38 | 42 | 23 | 24 | 29 | 31 | 36 | 32 | 22 | 20 | 18 | 16 | 17 | 12 | 16 | 23 | 29 | 22 | 15 | 16 |
| 17 | 1 | 1 | 2 | 1 | 2 | 4 | 6 | | 3 | 2 | 1 | 1 | 6 | 4 | 4 | 3 | 6 | 4 | 3 | 4 | 5 | 7 | 3 | 3 | 1 | 1 | 1 | | 1 | | 1 | 1 | | 2 | | 1 |
| 18 | | | | | | 1 | | | 1 | 1 | 5 | 5 | 2 | 2 | 1 | 3 | 6 | 4 | 7 | | 3 | | 1 | | 1 | 1 | | | 4 | | 1 | | | | | |
| 19 | 3 | | 1 | 3 | 1 | 2 | 1 | | | 1 | | | | | | | | | | | | | | | | | | 2 | | 2 | 1 | 3 | 2 | 3 | 7 | 3 |
| 20 | 17 | 11 | 2 | 3 | 6 | 4 | 3 | 2 | 4 | 5 | 4 | 6 | 3 | 3 | 7 | 1 | 7 | 5 | 5 | 6 | 2 | 5 | 4 | 1 | 6 | 9 | 6 | 4 | 15 | 8 | 10 | 12 | 16 | 31 | 36 | 27 |
| 21 | 38 | 24 | 15 | 11 | 11 | 9 | 8 | 9 | 16 | 14 | 10 | 11 | 20 | 24 | 23 | 15 | 17 | 26 | 33 | 22 | 20 | 11 | 9 | 21 | 8 | 18 | 21 | 24 | 19 | 30 | 31 | 42 | 51 | 40 | 28 | 27 |
| 22 | 13 | 13 | 6 | 4 | 4 | 7 | 5 | 3 | 5 | 5 | 6 | 6 | 9 | 8 | 14 | 26 | 29 | 21 | 25 | 28 | 18 | 10 | 26 | 26 | 25 | 16 | 16 | 17 | 17 | 21 | 16 | 17 | 14 | 5 | 10 | 21 |
| 23 | 3 | 3 | 3 | 4 | 1 | 2 | 5 | 3 | 3 | | 1 | 6 | 4 | 1 | 1 | 10 | 9 | 11 | 9 | 7 | 14 | 12 | 18 | 11 | 9 | 7 | 2 | 4 | 1 | 1 | 4 | 5 | 3 | 1 | 1 | 4 |
| 24 | 1 | 1 | | | 2 | 2 | 3 | 1 | 6 | 2 | 2 | 7 | 8 | 6 | 7 | 4 | 12 | 18 | 19 | 17 | 14 | 14 | 12 | 9 | 5 | 3 | 5 | 2 | 11 | 6 | 3 | 1 | 3 | 1 | | |
| 25 | | 1 | | | | 1 | | | | | | | | | | | | | | | | | | | | | | 2 | | | | | 1 | | | |
| 26 | | 2 | 1 | | | | | | | | | 1 | 2 | 1 | 1 | 1 | 1 | 2 | 1 | 3 | 1 | | 1 | 4 | 2 | 1 | 4 | 5 | 8 | 4 | 5 | 9 | 5 | 4 | 9 | 1 |
| 27 | 1 | 1 | | 1 | | | 2 | | | 1 | | 3 | | | 2 | | 1 | 2 | 2 | 1 | 3 | 2 | 5 | | 6 | 2 | 5 | 7 | 1 | 8 | 4 | 10 | 6 | 4 | 4 | 4 |
| 28 | 1 | | | | | | | | | | | | | | | | | | | 1 | 1 | 1 | 1 | | 1 | | | 1 | | | | | 1 | | | |
| 29 | | | | | | | | | | | | 1 | | 1 | | | | 2 | 2 | 1 | 1 | 2 | 4 | 2 | 2 | 1 | 2 | 1 | 1 | 3 | 1 | 1 | | | | |
| 30 | 3 | | 1 | 1 | | | | 1 | | | | | | 2 | | | 2 | | 1 | 2 | 2 | 3 | 6 | 2 | 2 | | | 2 | 1 | 1 | | 1 | | | | |
| 31 | | | | | | | | | | | | | | | | | | | | | 1 | | | | | | | | | | | | | | | |
| 32 | | | | | | | | | | | | | | | | | | | | | | 1 | | | | | 1 | | | | | | | | | |
| 33 | | | | | | | | | | | | | | | | | | | | | | 1 | | | | | | | | | | | | | | |
| 34 | | 1 | | | | | | | | | | | | | | | | | | | 1 | | | | | | | | | | | | | | | |
| 35 | | 1 | | | | | | | | | | | | | | | | | | | | | | | | | | | | | | | | | | |
| 36 | | 1 | | | | | | | | | | | | | | | | | | | | | | | | | | | | | | | | | | |

Table 2. *Random Directions of Residential Movement*, derived from Model 2. Each entry in the table represents the number of households with angles of movement falling in that sector.

| Region | Sector (10 degree) | | | | | | | | | | | | | | | | | | | | | | | | | | | | | | | | | | |
|---|---|---|---|---|---|---|---|---|---|---|---|---|---|---|---|---|---|---|---|---|---|---|---|---|---|---|---|---|---|---|---|---|---|---|---|
| | 1 | 2 | 3 | 4 | 5 | 6 | 7 | 8 | 9 | 10 | 11 | 12 | 13 | 14 | 15 | 16 | 17 | 18 | 19 | 20 | 21 | 22 | 23 | 24 | 25 | 26 | 27 | 28 | 29 | 30 | 31 | 32 | 33 | 34 | 35 | 36 |
| 1 | | | | 1 | 1 | 1 | | | | 1 | 1 | 1 | | | | 1 | | | 1 | 1 | | | | | | | 1 | 1 | | | | 1 | | | | 1 |
| 2 | | | | | | 2 | | 1 | | | | | | 1 | | | 2 | | | | | 1 | | | | | | | | | | | 1 | | | |
| 3 | | | | 1 | | | | | | | | 2 | | | 1 | | | | | | | | | | 1 | 1 | | | | | | | | | 2 | 1 |
| 4 | | | | | | | | | | | | | | | | | | | | | | | | | 1 | | | | | | | | | | | |
| 5 | 1 | | | | | | | | | | | | | | | | | | | | | | | | | | | | | | | | | | | |
| 6 | | | | 2 | | 1 | 2 | | | | 2 | | | | | 1 | | | 1 | | 1 | 1 | 2 | | | | 1 | | 1 | | | | | 1 | | |
| 7 | | | | | | | | | | | | | | | | | | | | | 1 | | | | | 1 | | | | | | | | | | |
| 8 | 1 | 3 | 3 | 1 | 2 | 1 | 2 | 2 | | | 2 | 2 | 3 | 5 | 2 | 1 | 2 | | 4 | 2 | 2 | 2 | | 5 | | 2 | 1 | 3 | 2 | 3 | 3 | 7 | | 1 | | 4 |
| 9 | 4 | 6 | 13 | 10 | 11 | 14 | 15 | 7 | 12 | 11 | 2 | 6 | 5 | 13 | 8 | 9 | 11 | 9 | 8 | 9 | 18 | 14 | 18 | 11 | 12 | 6 | 8 | 8 | 8 | 14 | 18 | 18 | 12 | 10 | 11 | 9 |
| 10 | 5 | 8 | 8 | 10 | 11 | 10 | 3 | 8 | 7 | 7 | 4 | 6 | 10 | 14 | 7 | 8 | 7 | 7 | 4 | 3 | 3 | 8 | 12 | 8 | 5 | 9 | 1 | 3 | 3 | 8 | 10 | 8 | 6 | 4 | 6 | 10 |
| 11 | 2 | | | 3 | 2 | 2 | 1 | | | | 1 | 1 | 2 | 2 | 2 | 3 | 1 | 3 | 2 | 2 | | | 1 | 3 | | | | 4 | | 1 | 1 | 1 | 2 | 3 | 1 | 1 |
| 12 | 2 | 6 | 4 | 3 | 4 | 4 | 5 | 1 | 2 | 1 | 1 | 6 | 3 | 6 | 3 | 2 | 7 | | 7 | 3 | 1 | 5 | 4 | 3 | 3 | 4 | | 1 | | 2 | 1 | 3 | 3 | 2 | 3 | 2 |
| 13 | | | 1 | 1 | 1 | 1 | | | | | 2 | | | | | | 1 | | | 1 | | | 1 | | | | | | | 1 | | | 2 | 1 | 2 | 2 |
| 14 | 4 | 5 | 3 | 2 | 5 | 3 | 6 | 5 | 2 | 1 | 4 | 6 | 7 | 9 | 4 | 1 | 3 | 4 | 2 | 8 | 4 | 10 | 6 | 6 | 5 | 3 | 5 | 2 | 3 | 5 | 3 | 6 | 7 | 5 | 7 | |
| 15 | 12 | 10 | 13 | 19 | 24 | 17 | 13 | 14 | 10 | 12 | 10 | 13 | 22 | 26 | 17 | 12 | 11 | 9 | 13 | 8 | 13 | 23 | 33 | 26 | 16 | 9 | 13 | 13 | 14 | 21 | 22 | 20 | 17 | 14 | 14 | 15 |
| 16 | 25 | 23 | 26 | 38 | 35 | 28 | 34 | 30 | 22 | 15 | 29 | 27 | 42 | 41 | 40 | 23 | 22 | 22 | 19 | 19 | 23 | 27 | 40 | 32 | 29 | 25 | 16 | 25 | 23 | 36 | 33 | 42 | 37 | 19 | 24 | 24 |
| 17 | 3 | 6 | 1 | 3 | 5 | 1 | | 2 | 5 | 1 | 3 | 2 | 1 | 8 | 3 | 4 | 2 | 2 | 1 | 3 | 1 | 4 | 2 | 4 | 2 | | 1 | 1 | 2 | 2 | 1 | 2 | 2 | 2 | 1 | 2 |
| 18 | 2 | 2 | 3 | 1 | | 1 | | 1 | | | | | 2 | 2 | 2 | 2 | | | 2 | | 3 | | 1 | 7 | 3 | | 1 | 3 | 2 | 1 | 3 | 2 | 1 | 1 | 1 | |
| 19 | 2 | | 2 | 2 | | | | 1 | 1 | | 1 | | 2 | 2 | 1 | 2 | 1 | 1 | 1 | | 3 | 2 | | 1 | | | 1 | | | 1 | 1 | 3 | 1 | 1 | 1 | 1 |
| 20 | 6 | 3 | 4 | 6 | 16 | 9 | 9 | 6 | 10 | 4 | 6 | 9 | 11 | 11 | 6 | 10 | 7 | 4 | 10 | 6 | 6 | 8 | 17 | 9 | 6 | 7 | 10 | 3 | 7 | 13 | 14 | 14 | 4 | 9 | 9 | 7 |
| 21 | 20 | 19 | 11 | 24 | 33 | 23 | 18 | 17 | 20 | 20 | 10 | 16 | 22 | 26 | 28 | 23 | 24 | 16 | 14 | 19 | 16 | 13 | 26 | 30 | 19 | 18 | 17 | 12 | 28 | 19 | 26 | 31 | 32 | 24 | 19 | 23 |
| 22 | 10 | 12 | 15 | 19 | 23 | 13 | 15 | 10 | 13 | 12 | 12 | 16 | 17 | 21 | 18 | 23 | 14 | 11 | 16 | 11 | 14 | 20 | 15 | 6 | 12 | 15 | 12 | 9 | 11 | 12 | 20 | 12 | 17 | 13 | 12 | 11 |
| 23 | 2 | 3 | 2 | 9 | 6 | 8 | 9 | 5 | 3 | 4 | 5 | 4 | 2 | 11 | 3 | 6 | 1 | 4 | 5 | 5 | 2 | 5 | 9 | 4 | 4 | 3 | 7 | 6 | 6 | 6 | 4 | 8 | 6 | 4 | 5 | 7 |
| 24 | 6 | 4 | 7 | 6 | 5 | 3 | 2 | 4 | 9 | 3 | 4 | 2 | 6 | 11 | 7 | 6 | 4 | 3 | 3 | 7 | 11 | 8 | 10 | 7 | 8 | 6 | 6 | 5 | 4 | 10 | 3 | 8 | 7 | 3 | 5 | 4 |
| 25 | | | | | | | 1 | | | | | | | 1 | | | 1 | | | 1 | | | | | | | | | | | | | | | 1 | |
| 26 | 2 | 2 | 1 | 3 | 3 | 1 | | 5 | 1 | 2 | 3 | 2 | 3 | 5 | 1 | 1 | | 1 | | | | | 5 | 3 | 4 | 1 | 1 | 4 | 4 | 1 | 3 | 9 | 3 | 5 | | |
| 27 | 1 | 1 | 2 | | 8 | 4 | 2 | 1 | | | 3 | 2 | 2 | 6 | 3 | 5 | 1 | 5 | 1 | 4 | 2 | 5 | 1 | 2 | 4 | 4 | 1 | | 3 | | 4 | 3 | 4 | 1 | 2 | 1 |
| 28 | 1 | | | | | | | 1 | 1 | | | | | | | | | | 1 | | 1 | 1 | | 1 | | | | 1 | | | | | | | | |
| 29 | | 1 | 1 | 2 | 2 | | | | | 1 | 2 | | 2 | 4 | | 1 | 1 | | | | 1 | | 2 | 2 | | 1 | | | | | | 2 | 2 | 1 | | |
| 30 | | 2 | 1 | | 2 | | | 1 | 3 | 1 | 2 | 1 | 1 | 1 | | | 1 | | | | 1 | 1 | | 1 | 2 | | | 1 | 1 | 2 | 1 | 2 | 2 | 1 | 1 | 1 |
| 31 | | | | | | 1 | | | | | | | | | | | | | | | | | | | | | | | | | | | | | | |
| 32 | | | | | | | | | | | | | | | | | | | | | | | | | | | | | | 1 | | | | 1 | | |
| 33 | | | | | | | | | 1 | | | | | | | | | | | | | | | | | | | | | | | | | | | |
| 34 | | | 1 | | | | | | | | | | | | | | | | | | | | | | | | | | | | | | 1 | | | |
| 35 | | | | | | | | | | | | | | | | | 1 | | | | | | | | | | | | | | | | | | | |
| 36 | | | | | | | | | | | | | | | 1 | | | | | | | | | | | | | | | | | | | | | |

for which there is no conflict. For example, region 26 has a maximum difference of 0.27 but region 15 has a maximum difference of only 0.13.

The differences between regions with biased directions of movement and the regions with a pattern of movement not basically different from random are further illustrated in a series of computer-calcomp plots of angles and distances of movement for households within the city. Six selected regions are reproduced here although all thirty-six regions were plotted (figures 6–11). Not all moves are plotted for each region. If there were fewer than fifty moves, all directions and distances of movements were plotted, but in the cases where there were more than fifty moves, a sample which preserved the overall patterns of the region was plotted. The difference in the originating region size shown in the graphs is merely a scale difference used for the calcomp plotter. To match the scale of the computer-drawn figures to figures 4 and 5, coordinates on the computer-drawn graphs need to be multiplied by a factor of ten.

Regions 8, 11, 16, 21, 26, and 29 were selected for reproduction here. Regions 8 and 11 (figures 6 and 7) show a bias towards the northeast and northwest, respectively. Although the angles of movement cover approximately 120°, most of the movements are concentrated within a smaller span. The regions north of the city center, regions 26 and 29 (figures 10 and 11), perhaps show a more marked bias in the movement patterns. It is worth noting that while region 11 (figure 7) has significant tests for both the twelve and thirty-six sector analyses, region 8 tests significantly for the thirty-six sector model only. Both regions 16 and 21 (figures 8 and 9) have all angles of movement represented. The pattern of movements fans out in what is best described as a star pattern. The calcomp plots serve another purpose in addition to that of further illustrating the differences between biased and random regions in terms of angles of moves made by households. It has been suggested that the movements of shorter distances may be essentially random, but that the longer-distance moves will be directionally biased (Adams 1969). Although no tests of this hypothesis were undertaken, the plots of angles of movement do not seem to support the hypothesis. It is possible that in the case of region 21, where more short-distance moves lie in a generally northeast-southwest direction, that a sectoral bias might be detected. However, even in this case the remainder of the moves cover approximately 70° in one direction and 90° in the other direction. Because this analysis of the directions of movement is graphically descriptive, it is more useful to test the second model of movement.

For the regions with five or more moves which did not fit the model of random moves, a test of the sector model was undertaken. Nine of the twenty-seven regions were identified as having a pattern of moves not significantly different from random on both the twelve sector and thirty-six sector tests. For the other eighteen regions, a simple and direct test of the sector model was carried out. Those sectors which made up approximately 50 per cent of all moves were identified (figure 12). If the sectors were contiguous and comprised no more than two totaling 60°, the region was allocated as fitting the sector model. Of the eighteen regions, fourteen fitted the requirements of the model (figure 12). The regions were 1, 2, 3, 6, 8, 11, 13, 18, 19, 25, 26, 28, 29, and 30. Some of the regions had only 46 or 48 per cent of the moves within the required two sectors but were considered as fitting the model because they fell close to the 50 per cent requirement and were considerably in excess of 50 per cent if a third sector was added. Almost all of the regions which did not fit the sector model were those with conflicting results on the tests of randomness; that is, they were significant for thirty-six sector test but not for the twelve sector test. One anomaly can be identified. In the case of region 12, the tests show the region as not significantly different from random, but the sector model also fits for the region. No adequate explanation for this apparent contradiction can be advanced at the present time, except to note that the maximum difference used in the Kolomogorov-Smirnov test is almost sufficient to make the region significantly different from random at the 0.05 level.

It might be argued that sectoral movement of households does not require that all moves (from an arbitrary zero point) be in one direction only. Thus, an investigation of the "mirror image" of the dominant sectors was undertaken, but the number of moves proved insufficient to alter the basic results; that is, none of the five regions for which the sector model was an inadequate fit had sufficient moves in the "mirror image" sectors to bring the percentage of moves between 0° and 60° to more than 50 per cent.

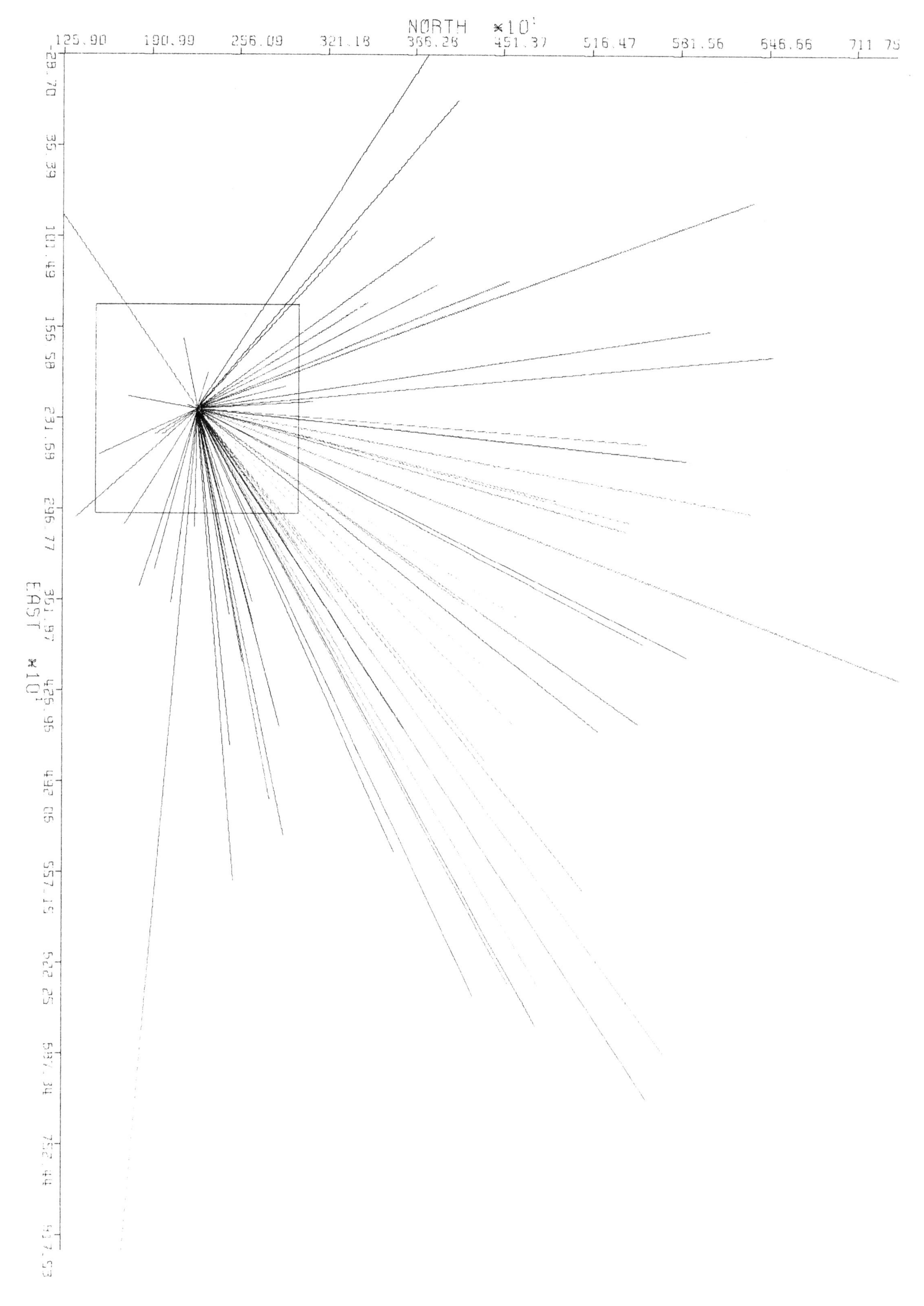


Figure 6. Region 8. Originating cell and directions and distances of observed moves.

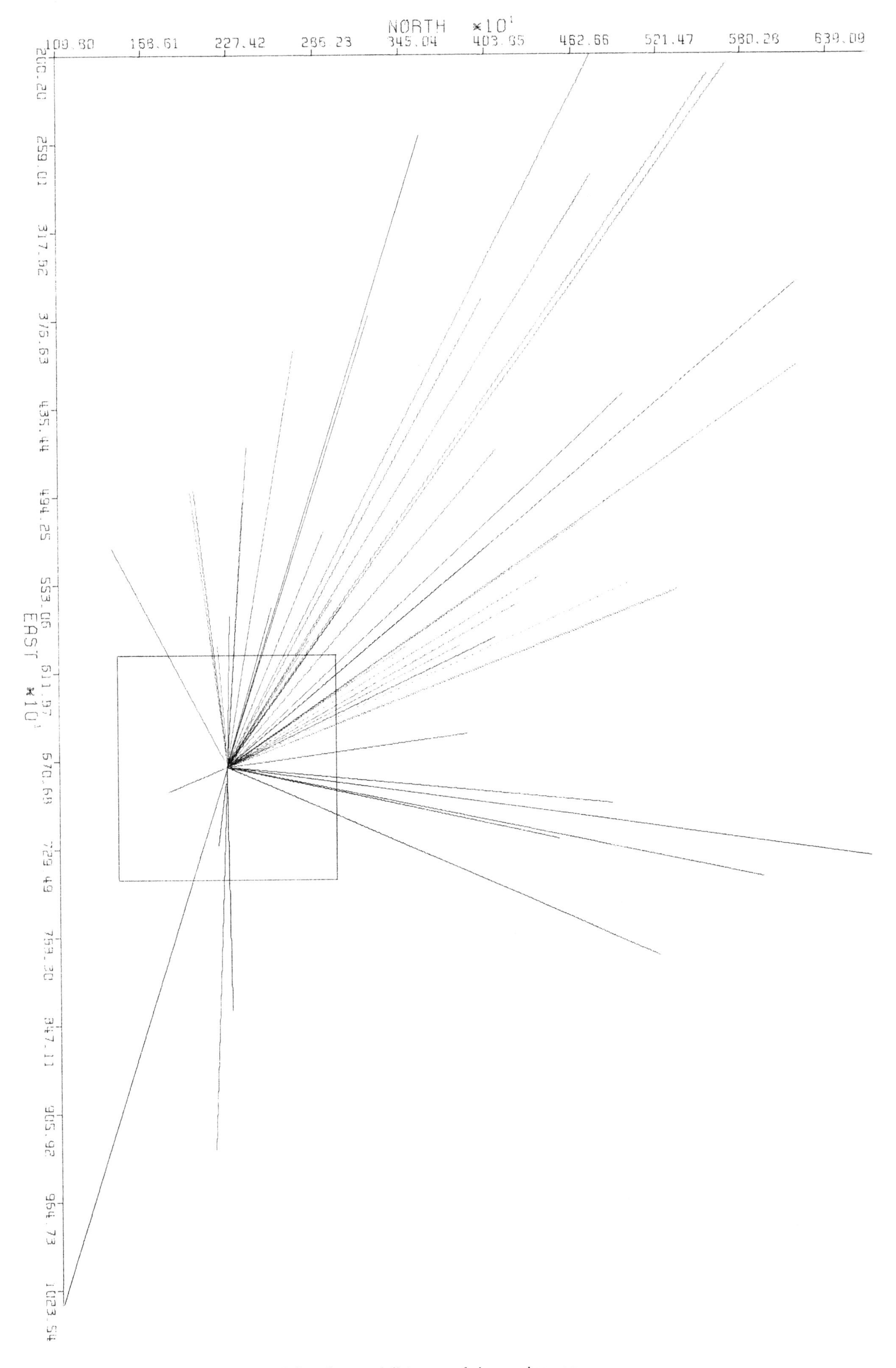


Figure 7. Region 11. Originating cell and directions and distances of observed moves.

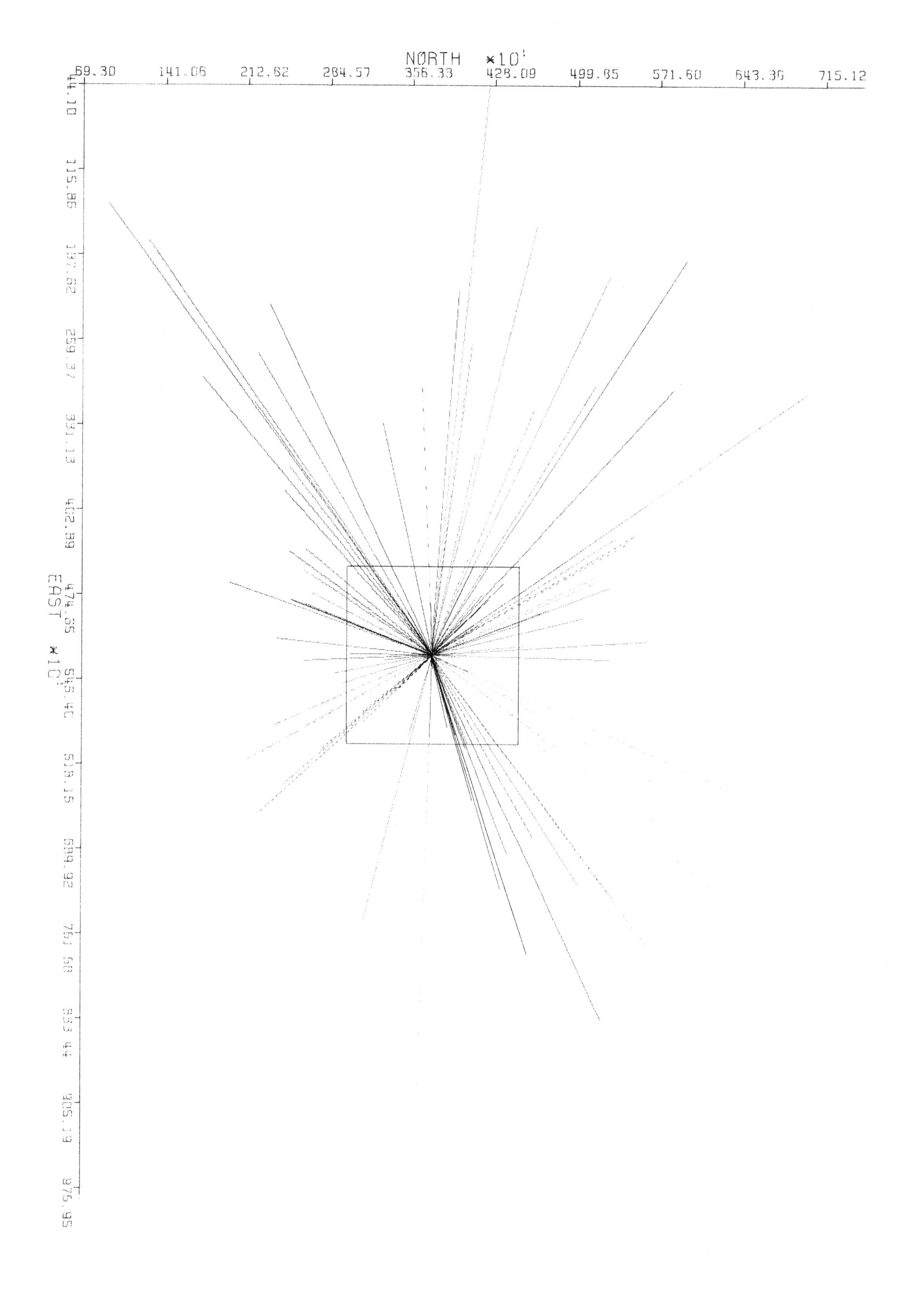


Figure 8. Region 16. Originating cell and directions and distances of observed moves.

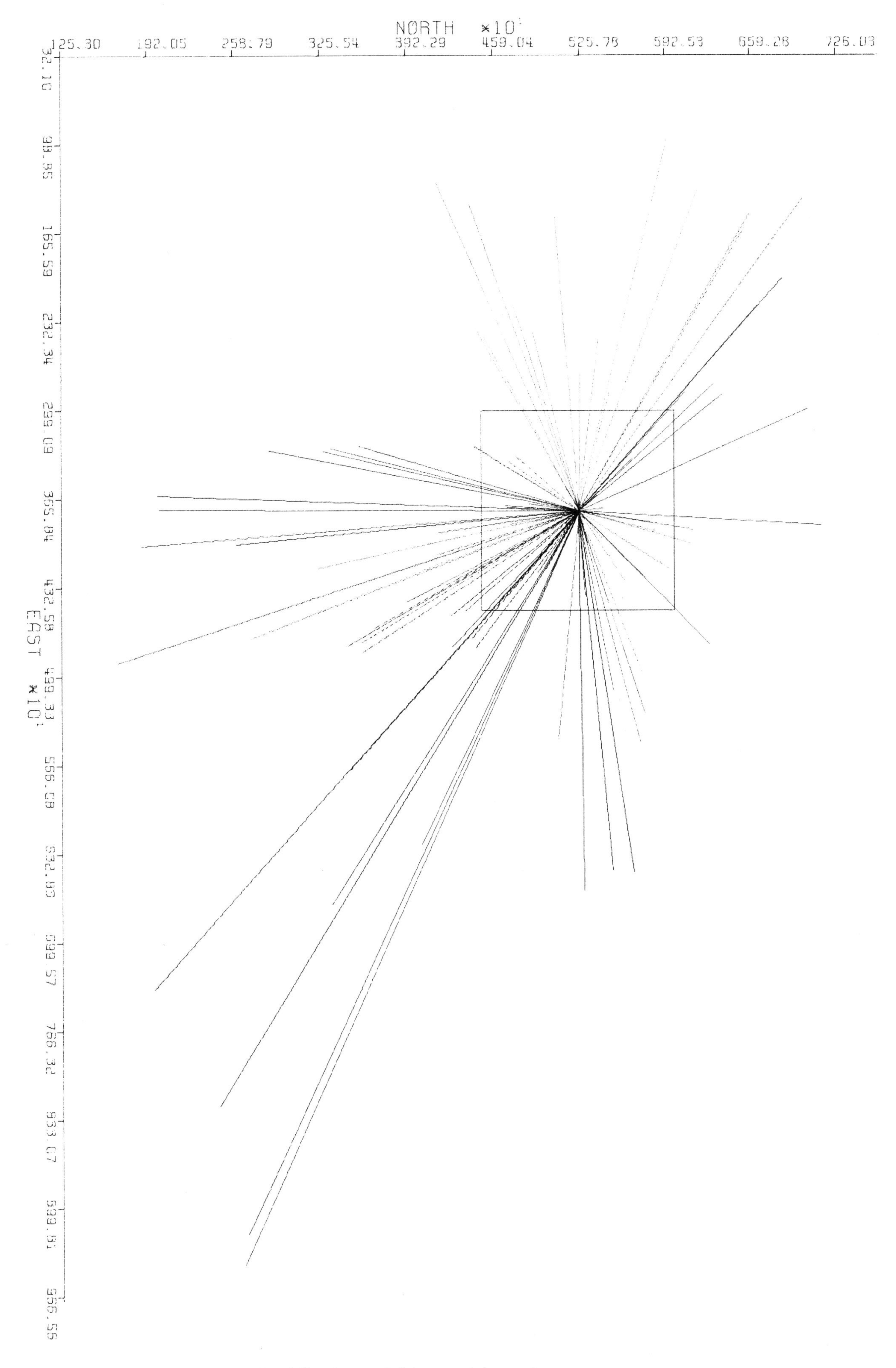


Figure 9. Region 21. Originating cell and directions and distances of observed moves.

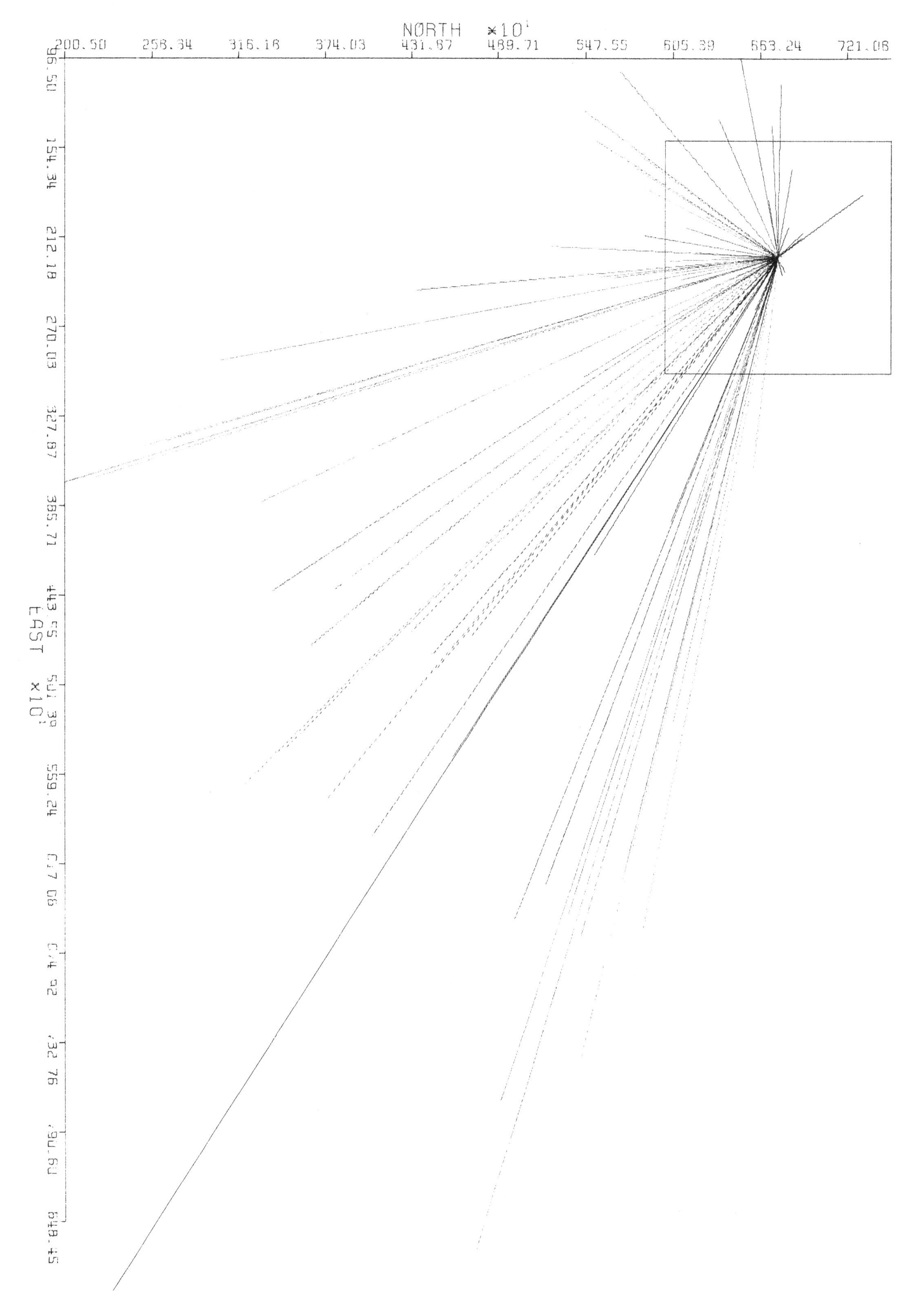


Figure 10. Region 26. Originating cell and directions and distances of observed moves.

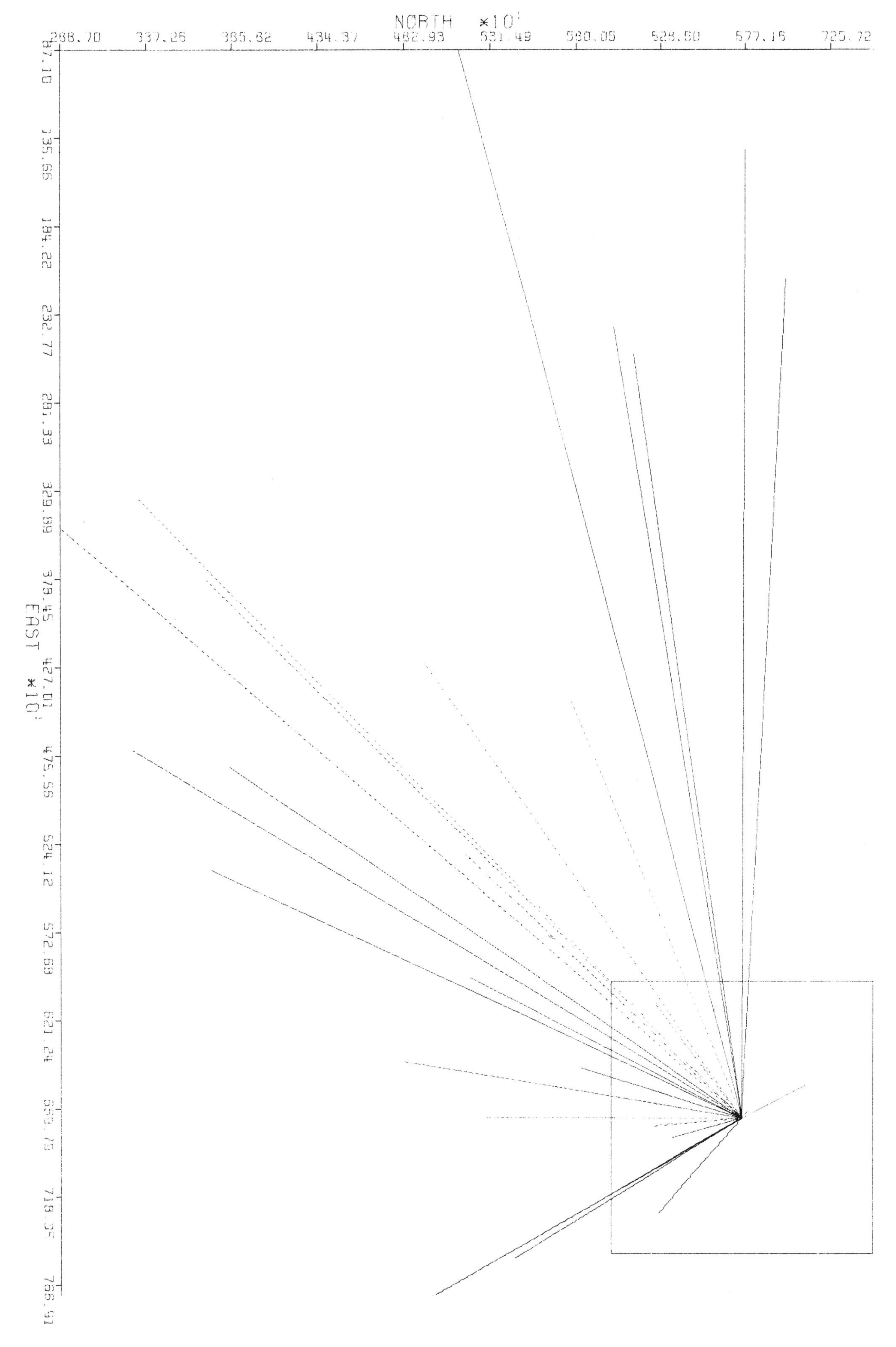


**Figure 11. Region 29. Originating cell and directions and distances of observed moves.**

Table 3. *Kolomogrov-Smirnov Tests of Differences between the Number of Random and Observed Household Moves, by Sector.*

| Region | 12 Sectors (30 degree) | 36 Sectors (10 degree) |
|---|---|---|
| 1 | 0.77** | 0.61** |
| 2 | 0.50** | 0.38** |
| 3 | 0.45** | 0.63** |
| 4 | 1.00** | 1.00** |
| 5 | 1.00** | 1.00** |
| 6 | 0.38 * | 0.43** |
| 7 | 0.50** | 1.00** |
| 8 | 0.28 | 0.45** |
| 9 | 0.22 | 0.37** |
| 10 | 0.23 | 0.28 * |
| 11 | 0.55** | 0.32 * |
| 12 | 0.36 | 0.21 |
| 13 | 0.71** | 0.37** |
| 14 | 0.32 | 0.32 * |
| 15 | 0.13 | 0.17 |
| 16 | 0.10 | 0.11 |
| 17 | 0.23 | 0.13 |
| 18 | 0.39 * | 0.37** |
| 19 | 0.57** | 0.40** |
| 20 | 0.20 | 0.27 * |
| 21 | 0.10 | 0.11 |
| 22 | 0.17 | 0.23 |
| 23 | 0.25 | 0.20 |
| 24 | 0.34 | 0.17 |
| 25 | 0.80** | 0.60** |
| 26 | 0.27 | 0.33** |
| 27 | 0.21 | 0.38** |
| 28 | 0.50** | 0.38** |
| 29 | 0.43** | 0.53** |
| 30 | 0.18 | 0.24 |
| 31 | 1.00** | 1.00** |
| 32 | 0.50** | 1.00** |
| 33 | 1.00** | 1.00** |
| 34 | 0.50** | 0.50** |
| 35 | 1.00** | 1.00** |
| 36 | 1.00** | 1.00** |

* Significant at the 0.05 level
** Significant at the 0.01 level

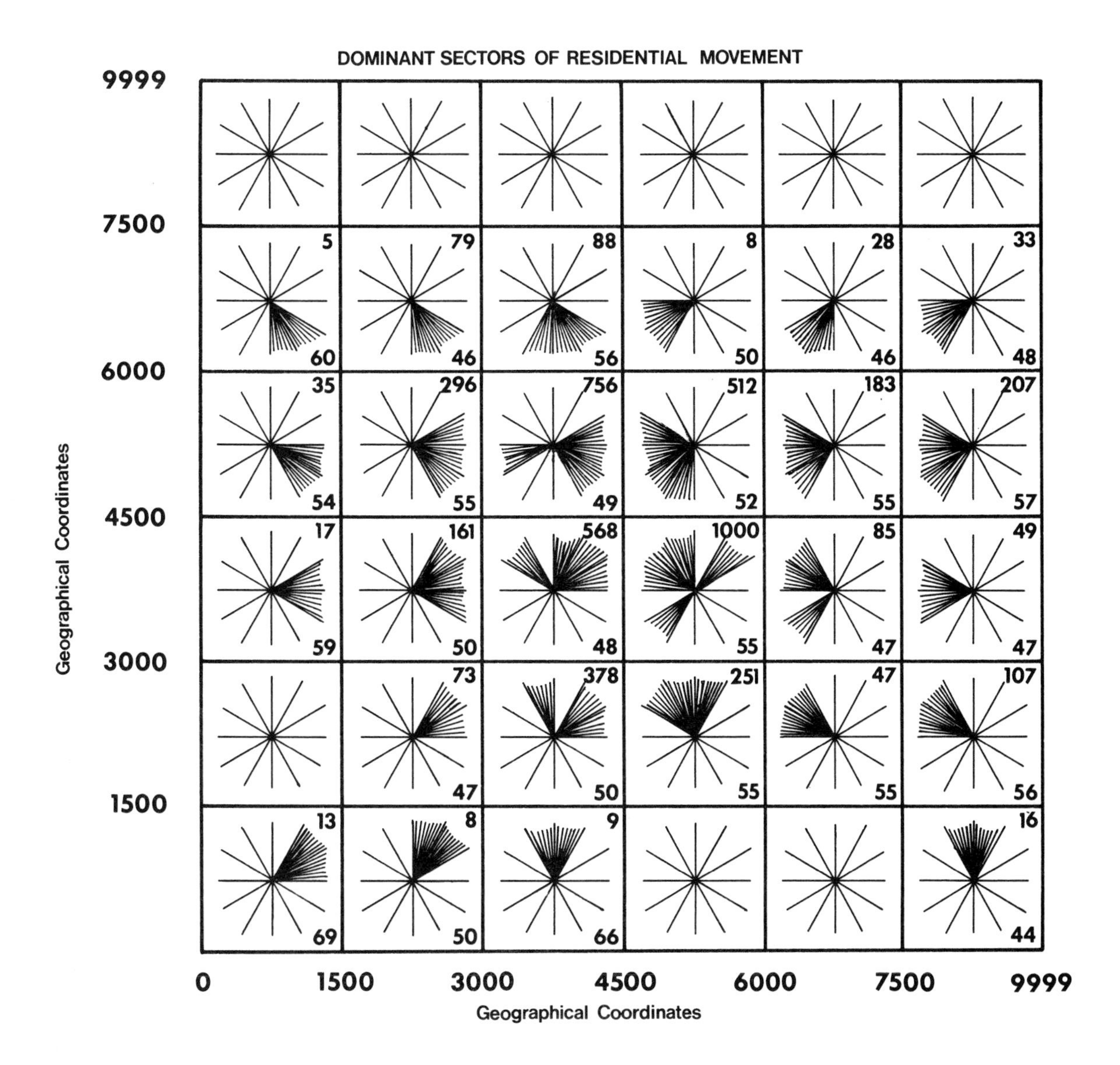


Figure 12. Dominant sectors of movement. The figure in the upper-right corner of each region indicates the total number of moves originating within the region, and the figure in the lower-right corner indicates the percentage of the total moves which occurred in the shaded sectors.

## CONCLUSIONS

The results of this investigation of directional bias cannot at this time be considered conclusive. For central regions of the city, the directions of movements cannot be differentiated from random. In the outer regions of the city, there is evidence that the movements of households are sectorally biased. However, two-thirds of the moves are accounted for by the nine nonbiased regions. This differentiation may reflect the pattern of available housing opportunities within the city. Within the central regions, choices of housing exist equally in all directions while in the outer regions the choice is more likely (because of greater availability) to be biased towards the central city. The results emphasize the importance of general housing opportunities within the city. Another suggestion is that the large number of apartment dwelling units in the central area may have a variety of family types with varying socioeconomic statuses who choose different areas of the city as their socioeconomic characteristics dictate. This differentiation of an inner and outer zone is also reflected in the portrayal of contact fields by Moore and Brown (1969).

Further testing of the patterns of movement and the extent of directional bias must await more detailed computer analysis, on the one hand, and detailed data sources, on the other. In particular, it will be necessary to establish the extent to which the economic sector is a dominant force in the relocation procedure. If, for example, the impact of the economic sector is not strong, then the neighborhood cells may be an important explanation and hence give rise to the more regular pattern of movements. While the directional-bias model cannot be completely rejected, the problem is more complicated than initially postulated. It is unfortunate not to be able to be more definite in the conclusions reached in this analysis, but the investigations of directional bias and structure have suggested some lines of inquiry worthy of further pursuit. Two avenues for further analysis can be suggested. First, a series of different regions both in size and arrangement may offer further ideas on the validity of the results presented here, and greater detail on the patterns of movement. Second, a variety of approaches involving moves of selected lengths may prove revealing despite the earlier comments on difference in length of move. On the other hand, analysis of urban structure and contact fields may be a more useful approach to the understanding of residential mobility. The identification of sectors within cities and the analysis of moves within these sectors may lead to insights into the patterns of movement. These kinds of investigations will require a more concentrated effort in detailed data collection than has heretofore been attempted. In addition, the implications for large urban centers, particularly in North America, must be explored. However, the in-depth presentation of models and data in this paper may allow other investigators to suggest more plausible models and explanations to deal with the analysis of structure and residential behavior.

## REFERENCES CITED


Adams, J. 1969. "Directional Bias in Intra-Urban Migration." *Economic Geography* 45:302–23.

Berry, B. J. L. 1965. "Internal Structure of the City." *Law and Contemporary Problems* 30:111–19.

Brown, L. A., and E. Moore. 1968. "The Intra-Urban Migration Process: An Actor Oriented Model." Department of Geography, University of Iowa.

Clark, W. A. V. 1969. "Information Flows and Intra-Urban Migration: An Empirical Analysis." *Proceedings, Association of American Geographers* 1:38–42.

———. 1970. "Measurement and Explanation in Intra-Urban Residential Mobility." *Tijdschrift voor Econ. en Soc. Geografie* 61: 49–57.

———, and Susan Stanman. 1970. "Intra-Urban Mobility and Indices of Residential Persistence." Department of Geography, University of California at Los Angeles.

Golledge, R. G. 1970. "Process Approaches to the Analysis of Human Spatial Behavior." Department of Geography, Ohio State University.

Gould, P. R. 1966. "On Mental Maps." Discussion Paper 9. *Michigan Inter-University Community of Mathematical Geographers.*

Lynch, K. 1960. *The Image of the City.* Cambridge, Mass.: The M.I.T. Press.

Marble, D. F., and J. D. Nystuen. 1963. "An Approach to the Direct Measurement of Community Mean Information Fields." *Papers and Proceedings of the Regional Science Association* 11:99–109.

Moore, E., and L. A. Brown. 1969. "Spatial Properties of Urban Contact Fields: an Empirical Analysis." Research Report No. 52. Department of Geography, Northwestern University.

Morrill, R. L., and F. R. Pitts. 1967. "Marriage, Migration and the Mean Information Field." *Annals Association of American Geographers* 57:339–61.

Murdie, R. 1969. "Factorial Ecology of Metropolitan Toronto, 1951–1961." Department of Geography, Research Paper No. 116. University of Chicago.

Orleans, P. 1968. "Sociology and Environmental Planning: a Scalar Perspective." Paper read to the American Sociological Association. Boston.

Rees, P. H. 1970. "Concepts of Social Space." In *Geographic perspectives in urban systems,* edited by B. J. L. Berry and F. E. Horton. Englewood Cliffs, N. J.: Prentice-Hall.

Rodwin, L. 1950. "The Theory of Residential Growth and Structure." *Appraisal Journal* 18:295–317.

Rossi, P. 1955. *Why Families Move.* Glencoe, Ill.: The Free Press.

Rushton, G. 1969. "Analysis of Spatial Behavior by Revealed Preference." *Annals Association of American Geographers* 59:391–400.

Siegel, S. 1956. *Nonparametric Statistics for the Behavioral Sciences.* New York: McGraw-Hill.

Simmons, J. W. 1968. "Changing Residence in the City." *Geographical Review* 58:622–51.

Stea, D. 1969. "The Measurement of Mental Maps: an Experimental Model for Studying Conceptual Spaces." In *Behavioral Problems in Geography: a Symposium,* edited by K. R. Cox and R. G. Golledge. Northwestern University, Studies in Geography No. 17.

Tilly, C. 1967. "Anthropology of the Town." *Habitat* 10:20–25.

# 2

# ON THE ARRANGEMENT AND CONCENTRATION OF POINTS IN THE PLANE

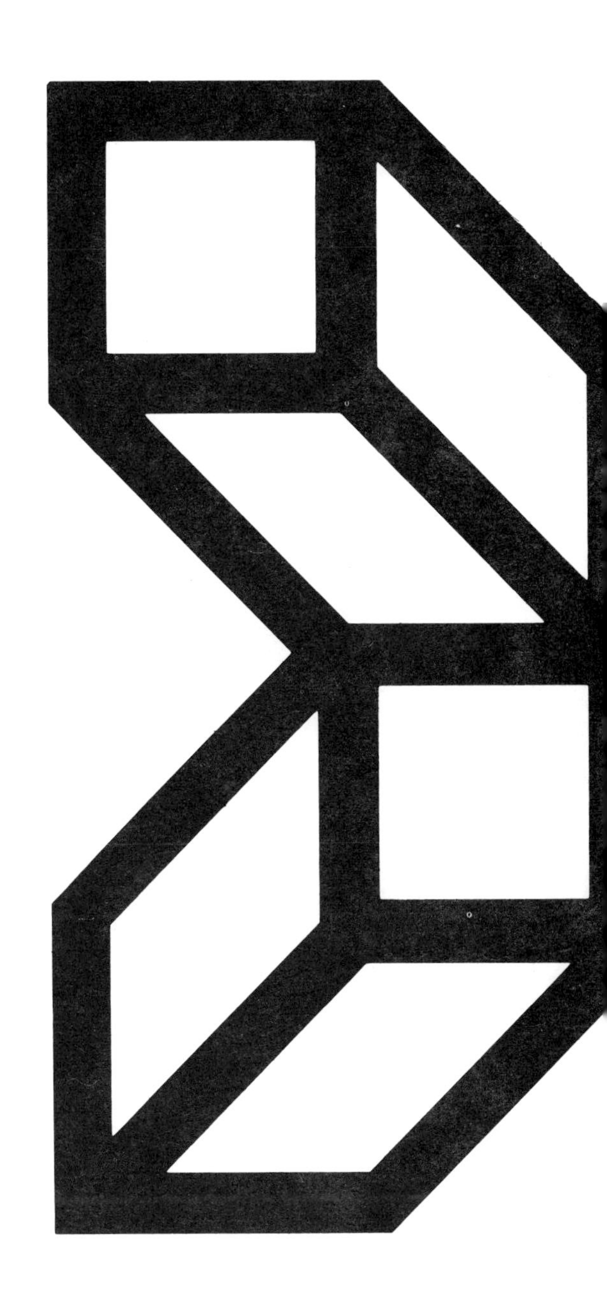

**Richard L. Morrill**
University of Washington

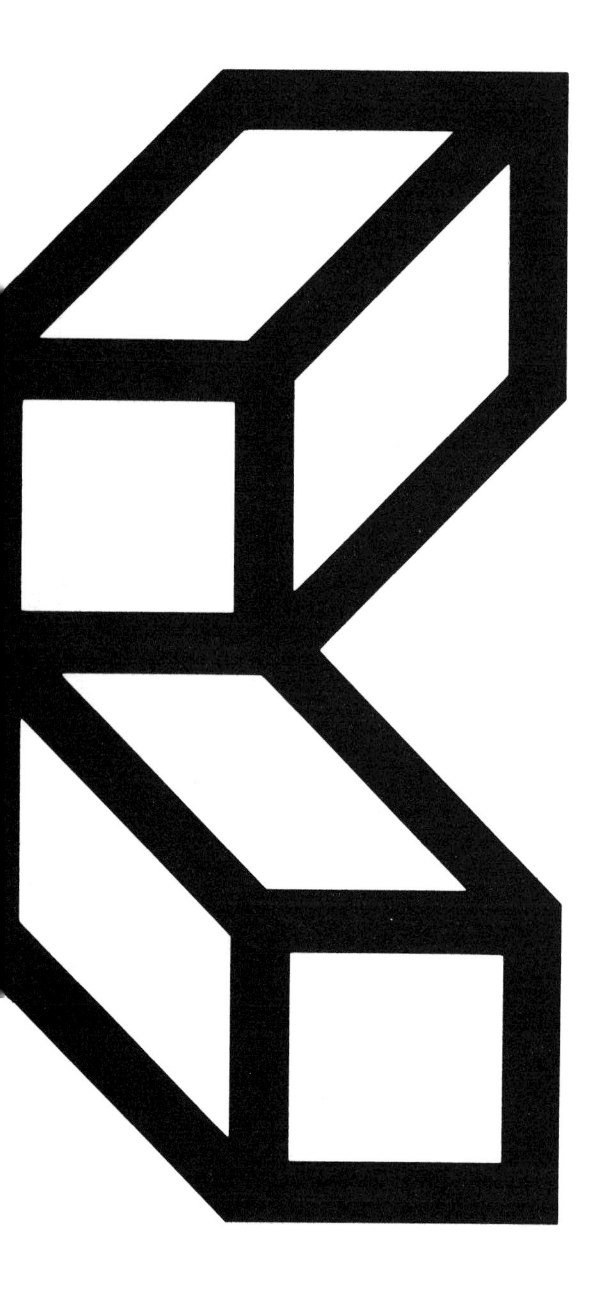

# On the Arrangement and Concentration of Points in the Plane

**ABSTRACT**

Attempts to measure randomness of points in a plane may lead to inconsistent and misleading results, because many empirical distributions, if not the majority of them, exhibit a concentration or clustering with respect to area, even though points or sets of points may show a random or even dispersed arrangement with respect to each other. Simple (and tentative) methods to deal with distributions of uneven density exist for situations, first, in which quadrat counts are available, and second, in which point locations are known. In the quadrat case, distributions more clustered than random are discussed. The measurement of randomness or dispersion (discontiguity) or clustering (contiguity) in the arrangement of sets of points is treated, whether or not the distribution of numbers of points in cells is random. In the discrete point case, means are provided for testing randomness or dispersion in the arrangement of points with respect to each other, even if the overall distribution of points in space is highly clustered. The test is based on the relationship between expected mean distance to nearest neighbor and mean distance from a center of a distribution. Finally, Poisson probabilities for unit areas of a circular normal distribution having exactly x occurrences are calculated.

Attempts to measure the "randomness" of points in a plane may lead to inconsistent and misleading results, simply because the possible variations in arrangement and concentration are such that the conditions necessary for the familiar tests of randomness become a very special case (Dacey 1964b, 1969).

Consider possible situations in which we have counts of objects in unit areas or choose to analyze the distribution in this manner. The Poisson distribution can give us the probabilities for the random expectation of numbers of points in cells of unit area, but spatially this does not get us very far. If the density of points over the surface is too variable, the distribution may test as clustered (concentrated), irrespective of the possible existence of a random or dispersed arrangement of cells. And even with respect to the special case of a fairly even density over the entire area, intriguing problems arise. In figures 2, and 3, the numbers of points in cells form a random distribution according to Poisson expectations. Yet in one case (2) the arrangement is strongly organized and dispersed, and in another (3), organized but concentrated (see King 1967). On the other hand, in figures 4, 5, and 6, the numbers of points in cells is decidedly nonrandom, yet again the arrangement of cells may be random, dispersed, or clustered. A simple method to deal with these problems is presented in part 1.

Consider also possible situations in which we know the exact location of individuals. The "nearest neighbor" (order distance) test is used to distinguish whether a distribution departs significantly from randomness in the direction of greater dispersion or greater clustering. But this is meaningful only if the density of points remains quite constant over the entire surface (as in figures 7 and 8). Otherwise results are misleading. Thus, in figure 9 and 10 we deliberately constructed the arrangement of points to be random and uniform in appearance (and by process), and yet, owing to the concentration of the major portion of points in a small part of the bounded area, the usual test would yield a clustered or concentrated result for both. It is true that the points are clustered with respect to the area, but they may be arranged randomly or uniformly with respect to each other. If the density of points over the area varies significantly, then the detection of randomness (or clustering or dispersion) of arrangement requires some alternative techniques. This is the task of part 2, the results of which are only tentative at this stage.

## PART I. CONCENTRATION AND ARRANGEMENT OF POINTS BY CELLS

The Poisson distribution gives the probabilities of events (objects, points) occurring in equal units of space (or time), if the spacing or timing has been produced by a random process—where each location is independent of every other (although not necessarily independent of a common reference point).

The familiar expression $e^{-\bar{x}}d^{x}/x!$ gives the probabilities of a unit of space (time) having exactly x occurrences, given that the mean number (density) of occurrences per unit area is $\bar{x}$. If we set the unit area such that $\bar{x} = 1$, then the expectations are given by $e^{-\bar{x}}/x!$ or

| x | Probability |
|---|---|
| 0 | 0.36788 |
| 1 | 0.36788 |
| 2 | 0.18394 |
| 3 | 0.06131 |
| 4 | 0.01533 |
| 5 | 0.00307 |
| 6 | 0.00051 |
| 7 | 0.000073 etc. |

For sample data, the values $\Sigma[(0 - E)^2/E]$ are distributed as chi-square for tests of significant departures from randomness.

*Distributions more dispersed than random.* If the observed frequency of 1 is larger than of 0 and most higher values, then the distribution is tending toward greater dispersion or uniformity. Obviously, maximum dispersion is simply

| x | Proportion |
|---|---|
| 0 | 0.0000 |
| 1 | 1.000 |
| 2 | 0.000 etc. |

i.e., each unit has exactly one occurrence.

Dacey (1964a) has discussed modified Poisson distributions which provide a description of how the probabilities will change systematically from random to maximum dispersion. For example, see table 1.

Table 1. *Change in P(x) from Random to Maximum Dispersion*

| Random | | 25% Dispersed | | 49% Dispersed | | 72% Dispersed | | 100% Dispersed | |
|---|---|---|---|---|---|---|---|---|---|
| x | prop. | x | prop. | x | prop. | x | prop. | x | prop. |
| 0 | 0.36788 | 0 | 0.30327 | 0 | 0.22225 | 0 | 0.12911 | 0 | 0.00000 |
| 1 | 0.36788 | 1 | 0.45490 | 1 | 0.58525 | 1 | 0.75097 | 1 | 1.000 |
| 2 | 0.18394 | 2 | 0.18554 | 2 | 0.16557 | 2 | 0.11119 | 2 | 0.00000 |
| 3 | 0.06131 | 3 | 0.04423 | 3 | 0.02434 | 3 | 0.00830 | | |
| 4 | 0.01533 | 4 | 0.00711 | 4 | 0.00241 | 4 | 0.00041 | | |
| 5 | 0.00307 | 5 | 0.00087 | 5 | 0.00018 | 5 | 0.000016 | | |
| 6 | 0.00051 | 6 | 0.00009 | 6 | 0.00001 | | | | |

It is readily apparent that these probabilities are sensitive or meaningful only if the density is rather even throughout the area. If the density is variable, then, even if the points are dispersed with respect to each other locally, the count of occurrences will yield proportions on the concentrated side of random.

*Distributions more clustered than random.* Most empirical distributions are likely to be "more clustered than random" simply due to density variations. Referring back to the original random expectations when the density, $\bar{x} = 1$, if the proportion with 1 is smaller, and of 0 larger, and of some higher values larger, then the distribution is tending toward greater clustering. Unfortunately, on this side the variability is infinite—maximum concentration is a simple function of n, the number of occurrences, since they may all (up to infinity) be in one unit area.

At this point we may utilize the simple variance measures of concentration (randomness or dispersal) as given by Barton and David (1959): $\Sigma n_i(x - \bar{x})^2/n - 1$ where $n_i$ is the number of cells with x occurrences. Now if we again set $\bar{x} = 1$, then the value of the measurement must be 0 for maximum dispersion, 1 for a random distribution, and $n - 1$ for maximum concentration. Thus:

| Dispersion | | | Concentration | | |
|---|---|---|---|---|---|
| $n_i$ | $x$ | $(x - \bar{x})^2$ | $n_i$ | $x$ | $(x - \bar{x})^2$ |
| 0 | 0 | 0 | 0 | 0 | 0 |
| $n_i$ | 1 | 0 | . . . . . . . . . . | | |
| 0 | 0 | 0 | $n_i$ | $x_i$ | $\Sigma(x_i - \bar{x})^2/n - 1 = n - 1$ |

Observe that in the spectrum from dispersion (0) to concentration ($n - 1$ approaching $\infty$), randomness, as traditionally measured in areal distributions, is very little different from maximum dispersion.

For a distribution which has a relatively even density throughout, the index of maximum concentration will be much less than $n - 1$. The maximum value, in fact, is a function of the number of cells for which we expect the density to be constant. The index value will with large n approach the quantity $(k - 1)^2/k$, where k is the number of cells for which density is constant. For example, if $k = 4$, that is, each set of four cells has 4 occurrences, but these are all concentrated in one out of the four—then the value of the measure will approach 2.25, if $k = 5$, the maximum index will be 3.2; if $k = 6$, 4.167; if $k = 10$, 8.1, and so forth. I have no good clues as to what value of k will insure relatively even density. This would have to be determined by the analyst in a particular study.

If our data points are for a random sample, then a variation of the earlier measure; namely: $\Sigma n_i(x - \bar{x})^2/x$ is distributed as chi-square at $n - 1$ degrees of freedom ($n =$ the number of objects) so we may test whether the distribution is significantly *nondispersed.* We may note that the value of chi-square, for maximum dispersion, must equal 0; and if random, it must equal n.

### *Randomness of the Arrangement of Occurrences by Cells*

The Poisson distribution discussed above and the

arguments thus far were concerned with the *numbers of points in cells,* but not with their *arrangement by cells.* It is quite possible, as we suggested in figures 1, 2, and 3, that while the distribution of numbers of points in cells is random, the arrangement of these sets of points need not be random. Similarly in figure 4, even though the number of points in cells is nonrandom, the arrangement of the sets may be random.

Whether or not the distribution of numbers of points in cells is random, figures 3 and 6 illustrate a situation of high contiguity of similar numbers (that is, cells with occurrences). There is a high positive correlation of values between adjacent cells. In other words, the arrangement of cells with objects shows concentration in the bounded area. Figures 2 and 5 illustrate the polar situation of high discontiguity of similar values—a tendency toward a high negative correlation of values between adjacent cells. Here the arrangement of cells with objects shows dispersion in the bounded area. In contrast, figures 1 and 4 illustrate a random or uncorrelated arrangement of numbers of points by cells.

Notions of *contiguity* have been devised to deal with these problems of arrangement. For example, King (1967) reports on methods by Dacey (1965) and others. For our present purposes, we prefer a very simple measure of contiguity, essentially the mean difference in occurrences between contiguous cells. The measure is: $c = (\Sigma d_{ij}/\bar{x})/n_c$, where $d_{ij}$ is the difference in numbers of points between any two contiguous cells, $\bar{x}$ is the mean number of points per cell; and $n_c$ is the number of contiguous edges within the bounded space; that is, the measure is the mean of the sum of differences between values over common boundaries. Note that if the mean number of points per cell is not preset at 1, then the sum of differences must be divided by this mean to form the measure.

With this measure, it is apparent that a value of 1 indicates randomness; a value less than 1, a clustering of like values; and a number greater than 1, a repelling of like values, or discontiguity.

In the examples shown in figures 1, 2, and 3, a random distribution of points in cells is chosen:

| $x$ | $n_i$ | $n_i x$ | $n_i(x - \bar{x})^2$ |
|---|---|---|---|
| 0 | 14 | 0 | 14 |
| 1 | 14 | 14 | 0 |
| 2 | 7 | 14 | 7 |
| 3 | 2 | 6 | 8 |
| 4 | 1 | 4 | 9 |
| | 38 | 38 | 38 |

In figure 1, the arrangement is random as well, since $c \simeq 1 \simeq 62/63$. In figure 2, the arrangement is dispersed (noncontiguous), since $c = 1.5 = 95/63$, while in figure 3, the arrangement is concentrated (contiguous), since

| | | | | | | |
|---|---|---|---|---|---|---|
| 1 | 0 | 0 | 1 | 2 | 0 | 2 |
| 1 | 2 | 1 | 1 | 0 | 0 | 1 |
| 2 | 0 | 3 | 2 | 0 | 1 | |
| 1 | 0 | 0 | 4 | 1 | 1 | |
| 3 | 1 | 1 | 0 | 0 | 2 | |
| 0 | 0 | 1 | 1 | 0 | 2 | |

RANDOM
$c \simeq 1$

Figure 1. Random Arrangement.

| | | | | | | |
|---|---|---|---|---|---|---|
| 1 | 0 | 1 | 0 | 1 | 2 | 0 |
| 0 | 3 | 0 | 1 | 0 | 1 | 2 |
| 1 | 0 | 1 | 0 | 1 | 2 | |
| 0 | 4 | 0 | 1 | 2 | 1 | |
| 1 | 0 | 1 | 3 | 0 | 2 | |
| 2 | 1 | 2 | 0 | 1 | 0 | |

DISPERSED
$c = 1.5$

Figure 2. Dispersed Arrangement.

Figures 1–3. Arrangement of Sets of Randomly Distributed Numbers of Points in Cells.

| | | | | | | |
|---|---|---|---|---|---|---|
| 1 | 2 | 2 | 2 | 2 | 3 | 4 |
| 1 | 1 | 1 | 2 | 2 | 2 | 3 |
| 1 | 1 | 1 | 1 | 1 | 1 | |
| 0 | 0 | 1 | 1 | 1 | 1 | |
| 0 | 0 | 0 | 0 | 0 | 0 | |
| 0 | 0 | 0 | 0 | 0 | 0 | |

CLUSTERED
$c = 0.3$

Figure 3. Clustered Arrangement.

| | | | | |
|---|---|---|---|---|
| 2 | 0 | 2 | 0 | 0 |
| 0 | 3 | 0 | 3 | 0 |
| 2 | 0 | 2 | 0 | 0 |
| 0 | 3 | 0 | 3 | 0 |
| 2 | 0 | 3 | 0 | 0 |

DISPERSED
$c^* = 1.64$

Figure 5. Dispersed Arrangement.

| | | | | |
|---|---|---|---|---|
| 0 | 0 | 3 | 2 | 0 |
| 0 | 0 | 3 | 0 | 0 |
| 0 | 2 | 0 | 0 | 2 |
| 2 | 3 | 2 | 0 | 3 |
| 3 | 0 | 0 | 0 | 0 |

RANDOM
$c^* = 1$

Figure 4. Random Arrangement.

| | | | | |
|---|---|---|---|---|
| 0 | 0 | 0 | 0 | 0 |
| 0 | 0 | 0 | 0 | 0 |
| 0 | 0 | 0 | 0 | 0 |
| 2 | 2 | 2 | 2 | 2 |
| 3 | 3 | 3 | 3 | 3 |

CLUSTERED
$c^* = 0.29$

Figure 6. Clustered Arrangement.

Figures 4–6. Arrangement of Sets of Non-Randomly Distributed Numbers of Points in Cells.

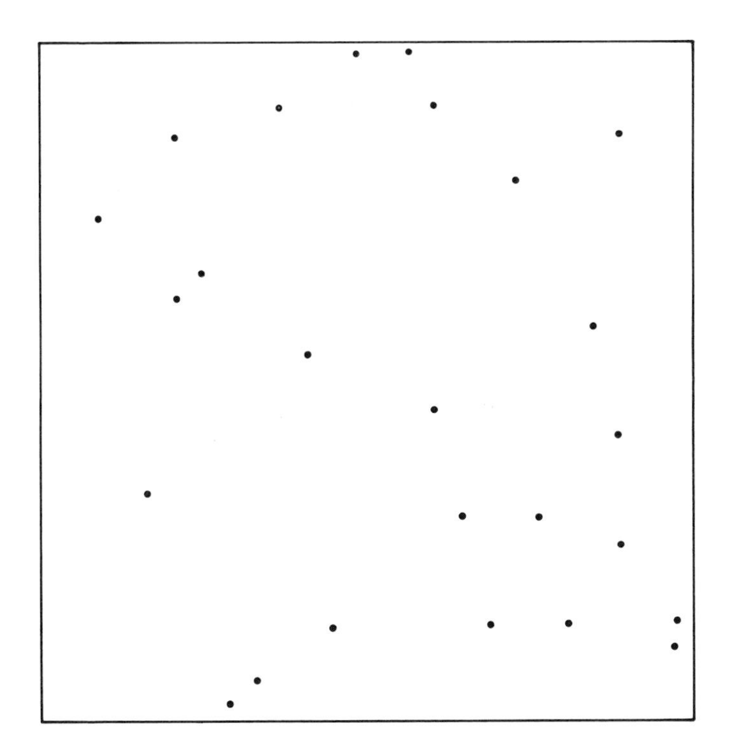

Figure 7. Randomly Generated Distribution of Points in the Plane ("Even" Density).
Area = 25
Density = 1
$\bar{r}_a = \bar{r}_e = 0.5$
$\bar{d}_a = \bar{d}_t = 2$

Figure 8. Regular Dispersed Arrangement of Points in the Plane.
Area = 25
Density = 1
$\bar{r}_a = 1$
$\bar{r}_e = 0.5$
$\bar{d}_a = \bar{d}_t = 2$

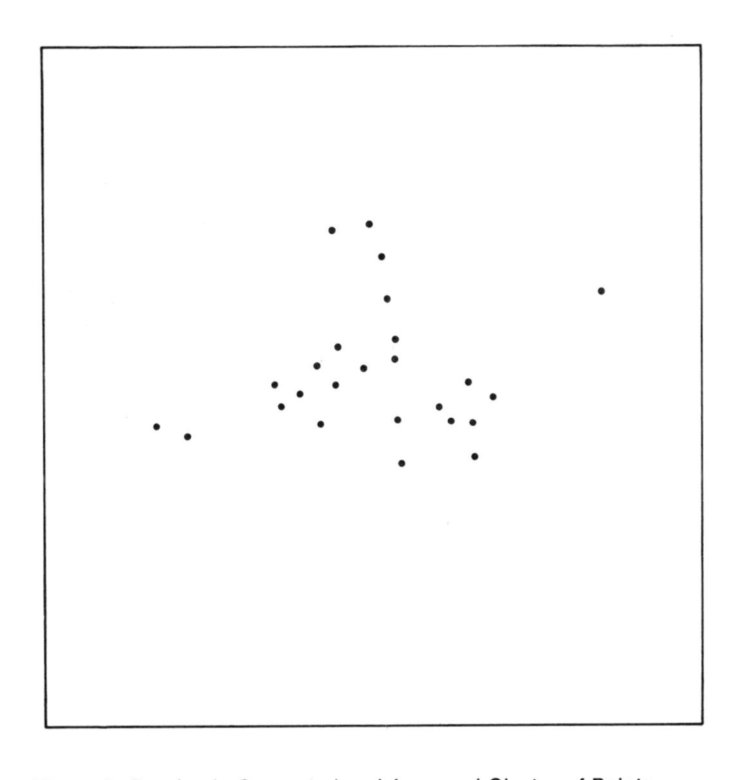

Figure 9. Randomly Generated and Arranged Cluster of Points According to Circular Normal Probabilities.
Area = 25
Density = 1
$\bar{r}_e = 0.5$
$\bar{r}_a = 0.23$
$\bar{d}_a = 0.8$
$\bar{r}_e^* = 0.2$

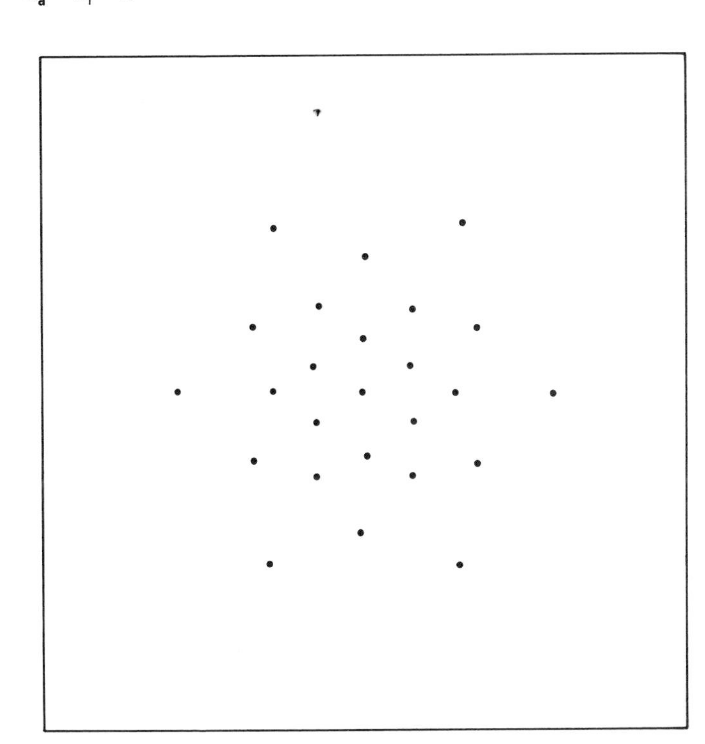

Figure 10. Dispersed Arrangement of Points in a Cluster Satisfying the Circular Normal Density Parameters.
Area = 25
Density = 1
$\bar{r}_e = 0.5$
$\bar{r}_a = 0.44$
$\bar{d}_a = 0.8$
$\bar{r}_e^* = 0.2$

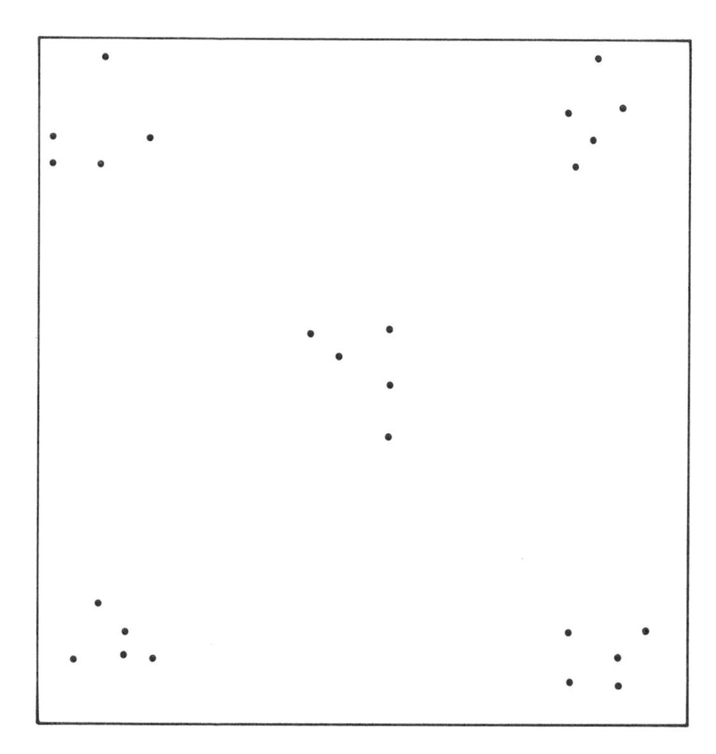

Figure 11. Clusters of Randomly Arranged Points Concentrated in 20% of the Area.

$c = 0.3 = 19/63$. Note that a regular gradient pattern is associated with maximum contiguity, a regular "waffle" density pattern—peaks and pits—with maximum discontiguity.

Similarly, for the examples in figures 4, 5, and 6, we derive a significantly nonrandom distribution of numbers of points in cells (25). The value of the earlier index is: $\Sigma n_i(x - \bar{x})^2/n - 1 = 40/24 = 1.67$. Zero and 3 frequencies are overrepresented; 1 is underrepresented. Unfortunately, the contiguity measure, as such, is affected by a departure from randomness in numbers of points in cells. A high concentration will tend to produce high values of the contiguity measure, whatever the pattern. To correct for this nonrandomness, the contiguity measure must be divided by the square root of the randomness measure (since the latter is based on squares of deviations). Thus: $c^* = c/\sqrt{\Sigma n_i(x - \bar{x})^2/n - 1}$. In our example, the uncorrected contiguity measures must be divided by $\sqrt{1.67}$, or 1.29. In figure 4, the arrangement is random, since $c^* = 1 = 1.29/1.29 = 51/40/1.29$. In figure 5, the arrangement is dispersed (non-contiguous) since $c^* = 1.64 = 2.1/1.29 = 85/40/1.29$. In figure 6, the arrangement is concentrated (contiguous) since $c^* = 0.29 = 0.37/1.29 = 15/40/1.29$.

Cases 4 and 5 are of particular importance; analysis of many map distributions by number of points in cells might well reveal a nonrandom concentration or clustering. Yet the clusters themselves might have a random arrangement or even a dispersed one, which may be of great theoretical interest. Thus, such a contiguity test should be a component part of a quadrat analysis of areal distributions.

The question of whether a departure from randomness of arrangement (degree of contiguity) is significant may be reduced to a simple test of whether an observed mean difference for a given sample size might really be the same as the expected mean difference, if the arrangement is random.

If we set $\bar{x} = 1$, as in these examples, then the expected mean difference is found to be 1, and if $\bar{x} \neq 1$, then the expected mean difference will equal $\bar{x}$ in a random arrangement. The standard deviation of the differences will also equal 1, if $\bar{x} = 1$, and $\bar{x}$, if it does not. Therefore, the standard error of the mean difference is simply $s_c = 1/\sqrt{n}$, if $\bar{x} = 1$; or $s_c = \bar{x}/\sqrt{n}$, if it does not. Thus for $\bar{x} = 1$, the test $\sqrt{n}|c - 1| = t$ is appropriate. In the examples:

$1: t = \sqrt{38}|1 - 1| = 0$ (random)
$2: t = \sqrt{38}|1.5 - 1| = 3$ (nonrandom, dispersed)
$3: t = \sqrt{38}|0.3 - 1| = 4.3$ (nonrandom, clustered)
$4: t = \sqrt{25}|1 - 1| = 0$ (random)
$5: t = \sqrt{25}|1.64 - 1| = 3.55$ (nonrandom, dispersed)
$6: t = \sqrt{25}|0.29 - 1| = 3.2$ (nonrandom, clustered)

The contiguity measure, with adjustment for nonrandomness (concentration) in numbers of points in cells, may be applied to distributions of highly variable density. For example, it was used experimentally with a circular normal distribution (generated by random numbers; figure 9) and its randomness of arrangement was recognized, despite its concentration in space. But application to a rigidly geometric and dispersed pattern (figure 10), satisfying the same density parameters, yielded not a discontiguous value (less than 1) but a slight degree of contiguity (clustering), because, with the concentration in space, the regularity of arrangement resulted in fairly high contiguity. Therefore, we must conclude that for highly concentrated, single-centered distributions, the contiguity test is not sensitive to arrangement of individual points with relation to each other, but it remains useful at the level of arrangement of sets of points.

## PART II. CONCENTRATION AND ARRANGEMENT OF POINTS IN A PLANE

In the introduction, we observed that the distance statistics (as nearest neighbor) test of randomness is meaningful only with respect to a distribution of reasonably even density. Otherwise, local dispersion and even randomness in the arrangement of points to each other will be masked by the marked apparent clustering due to density variation (or concentration of points within the bounded area).

In part 1 we noted that the contiguity measure, applied to counts in cells, was able to distinguish random, dispersed, or clustered arrangements of sets of points in cells, even in the presence of density variation (concentration in space), but was not able to distinguish randomness or nonrandomness in the patterns of individual points.

In this section we suggest a fairly simple method for identifying possible randomness or dispersion (uniformity) of arrangement within what is a concentrated distribution with respect to the total area.[1]

In a random distribution of fairly constant density (as figure 7), the expected distance to the nearest neighbor in point is: $0.5\sqrt{e}$ where e is density. Therefore, if we set the density $e = 1$, then $\bar{r}_e = 0.5$. Similarly, a density of 1 implies also a distance between neighbors of 1 for a regular (rectilinear) lattice (figure 8). There arises the question of the expected mean distance to nearest neighbor in a distribution which is random with respect to the arrangement of points to each other, but concentrated with respect to territory. We propose a tentative solution to certain cases of this problem.

Now the *mean distance* from the center of a distribution (again of fairly constant density) is a function of the *area*, such that $\bar{d} = \sqrt{A/2\pi}$ (A = area) irrespective

1. For related discussions, see Dacey (1968 and 1969).

of density and irrespective of whether the arrangement is random, clustered, or uniform locally.

For a given area, and for a continuous (that is, even or constantly changing density) distribution, there is a *constant* relation between $\bar{r}_e$, the expected mean distance to nearest neighbor, if the arrangement is random, and $\bar{d}$, the mean distance from the center of a distribution to all the points of the distribution. For example, with a density e = 1

| $\bar{r}$ | $\bar{d}$ | A |
|---|---|---|
| 0.5 | 1 | 6.28 |
| 0.5 | 2 | 25.1 |
| 0.5 | 3 | 56.5 |
| 0.5 | 4 | 100.5 |
| 0.5 | 5 | 157 |

Thus, for an area (and n) = 100, the ratio of $\bar{r}$ to $\bar{d}$ is 1:8. As such, this is not interesting, but of more significance is the fact that these ratios remain constant for some distributions that are not of even density, namely continuous but single-centered distributions; for example, the circular normal, or any other single-peaked or continuously changing distribution. It may be deduced that for these continuous distributions, the ratio of r to d reduces simply to $1.25/\sqrt{n}$.

For continuous distributions which, however, are not of even density, $\bar{d}_a$, the actual mean distance from the center to points must be calculated by finding the bivariate median, which, by definition is the point which minimizes the total distance:[2] min $\Sigma r$.

We may therefore work backwards from the calculated mean distance from the center, which is not affected by considerations of local arrangement (randomness or nonrandomness) and utilize the ratios to estimate $\bar{r}_e$, the expected mean distance to nearest neighbor, if the arrangement were random.

For example, in a distribution where e = 1, and A and n = 100, we would know that if the distribution were of even density and the arrangement random, that $\bar{r}_e = 0.5$ and $\bar{d} = 4$. But suppose the distribution is centered and we find that $\bar{d}_a = 2$, not 4. Then we can estimate that the $\bar{r}_e^*$ the corrected mean distance to nearest neighbor, for a random arrangement within a cluster, should be 0.25, not the 0.5 estimated from knowledge of density alone. Then the tests of whether $\bar{r}_a$, the actual mean distance to nearest neighbor, is significantly different from $\bar{r}_e$, can be made as usual.

The corrected expected mean distance to nearest neighbor is then: $\bar{r}_e^* = \bar{r}_e(\bar{d}_a/\bar{d}_t)$ where $\bar{r}_e$ is the expected mean distance based on density, $\bar{d}_t$ is the "territorial" mean distance (that is, based on the area), and $d_a$ is the actual calculated distance from the center to the points of the distribution.

In a randomly (but here, imperfectly) generated cir-

2. If a computer program to find the bivariate median is not available, only a very small error will be introduced by using the bivariate mean (center of gravity) instead.

cular normal distribution, figure 9 for example, we find that $r_a = 0.23$, the usual $r_e = 0.5$, which incorrectly suggests a nonrandom clustered arrangement, whereas based on the measured mean distance $d_a = 0.8$, $r_e^* = 0.2$ (since at $n = 25$, $r_e : d_e$ is 1:4).[3] The observed $r_a = 0.23$ is not significantly different from the theoretical $r_e^* = 0.21$, indicating a random arrangement within the clustered distribution.

In contrast, we find that $r_a = 0.44$, when compared to $r_e = 0.5$, gives a misimpression of randomness (or even clustering); but if it is compared to $r_e = 0.21$, then $r_a = 0.44$ correctly identifies the dispersed arrangement within the cluster.

In these examples, observe that the actual mean $d_a$ was 0.8, while the expected mean distance for an even distribution of that mean density would have been 2. Thus, we can utilize the ratio of territorial mean distance (here 2) to actual mean distance (0.8), or 2.5:1, as an indication of the degree of concentration in a single-centered cluster.

It is also quite possible, however, that a distribution may be multiple-centered, as one with large empty spaces between clusters (figure 11), and that the actual mean distance $d_a$ may well exceed $d_t$, thus indicating the degree of peripheral dispersion. Obviously the ratios of r to d will not be predictable in a multicentered or very irregular distribution. One way to deal with this problem (and a valuable description in itself of a distribution) is to utilize a program (such as SYMAP) to find the "peaks" and the "valleys" of the distribution; that is, the boundaries between major clusters, and then test each cluster separately for randomness of arrangement. If the clusters are easily identifiable, the grand mean distances from the center of each cluster to its members will permit the same procedures described above.

Another possible approach is to adjust the measurement of area (and consequently the density), but this is a far from simple task. One conservative method is by reference to Poisson expectations. Recall that at a density of 1, counts of points in cells, if the distribution is random, should be as indicated below. Now suppose we find that our actual distribution of proportions is rather different:

| | Theoretical | Actual |
|---|---|---|
| 0 | 0.368 | 0.80 |
| 1 | 0.368 | 0.00 |
| 2 | 0.184 | 0.00 |
| 3 | 0.061 | 0.00 |
| 4 | 0.015 | 0.00 |
| 5 | 0.003 | 0.20 |

We can't claim the entire 80 per cent of cells that are empty as not part of the meaningful area, since in a random distribution 37 per cent of the cells are expected

3. In the case of a theoretical distribution, $r_e^*$ can be estimated directly (Neft 1966), since in a circular normal distribution, $d = 0.833\ s_r$; and since in our example $r:d = 1:4$ and $e = 1$, then $r_e^* = 0.208$.

to have 0 occurrences. However, it is reasonable to claim the excess of 0 area over the random expectations, in this sample, 43 per cent. Here the area would be reduced 43 per cent, effectively raising the density to 1.76 and reducing $r_e$ to 0.375.

*Poisson Probabilities for Distributions More Clustered Than Random*

The above example also raises the possibility of finding the probabilities for occurrences in cells of specific clustered distributions, such as the circular normal, as a basis for comparison with sample clustered distributions. "Goodness of fit" chi-square comparisons of proportions in standard distance bands can test the density parameters but not the local arrangement of points.

For the circular normal case we may calculate the probabilities from:

$$\int_0^{\beta} \frac{r d^x e^{-d}}{x!} dr \text{ where } d = \alpha e^{-\beta r^2}$$

[in this case, $\alpha = 9$, $\beta = 3$] and $\alpha$ is such that $d = 1$; $\beta$ controls the extent of the distribution; r (radial distance) is measured in $s_r$ units, and the summation is over r for each value of x. At a given density, $d^x e^{-d}/x!$ gives the expected proportions of x occurrences in cells. But d (density) is a continuously changing function of r, and the area for which the factorial expectations apply is a linearly increasing function of r. Thus we may also write for each value of x

$$\int_0^{\beta} \frac{r(\alpha e^{-\beta r^2})^x e^{-(\alpha e^{-\beta r^2})} dr}{x!}.$$

The tentative calculations, subject to revision, are given in table 2.

A distribution which follows the density parameters of the circular normal but has lower representations of 0, and higher frequencies of 1, 2, and 3 occurrences indicates a tendency toward dispersion; while higher representations of 0, with lower frequencies of 1 and 2 but higher frequencies of certain higher numbers of occurrences (between 4 and 18), indicate a tendency toward nonrandom clustering within the overall cluster! However, the nearest-neighbor method discussed above is simpler and more sensitive than this procedure.

Table 2. *Probability of a Unit Area Containing x Occurrences* for a Circular Normal Distribution of Mean Density of One, Area of $(3s_r)^2$

| | | Of | Which | Within |
|---|---|---|---|---|
| x | Total | 0 to $1s_r$ | 1 to $2s_r$ | 2 to $3s_r$ |
| 0 | 0.68648 | 0.00095 | 0.14727 | 0.53826 |
| 1 | 0.11930 | 0.00394 | 0.09868 | 0.01668 |
| 2 | 0.05387 | 0.00854 | 0.04473 | 0.00060 |
| 3 | 0.03592 | 0.01288 | 0.02302 | 0.00002 |
| 4 | 0.02687 | 0.01552 | 0.01135 | 0.55556 |
| 5 | 0.02074 | 0.01565 | 0.00509 | |
| 6 | 0.01624 | 0.01416 | 0.00208 | |
| 7 | 0.01252 | 0.01175 | 0.00077 | |
| 8 | 0.00936 | 0.00910 | 0.00026 | |
| 9 | 0.00669 | 0.00661 | 0.00008 | |
| 10 | 0.00476 | same (x = 10–18) | 0.33333 | |
| 11 | 0.00308 | | | |
| 12 | 0.00187 | | | |
| 13 | 0.00108 | | | |
| 14 | 0.00065 | | | |
| 15 | 0.00033 | | | |
| 16 | 0.00015 | | | |
| 17 | 0.00006 | | | |
| 18 | 0.00003 | | | |
| | 0.11111 | | | |

## REFERENCES CITED


Barton, D. E., and F. N. David. 1959. "Contagious Occupancy." *Journal of the Royal Statistical Society* 21:120–33.

Dacey, Michael F. 1964a. "Modified Poisson Probability Law for Point Pattern More Regular than Random." *Annals of the Association of American Geographers* 54:559–65.

———. 1964b. "Two-dimensional Random Point Patterns." *Papers of the Regional Science Association* 13:41–55.

———. 1965. "A Review of Measures of Contiguity for Two and K-color Maps." *Technical Paper 2,* Office of Naval Research NONR-1228(33).

———. 1968. "Some Properties of Clustered Point Processes." Mimeographed. Department of Geography, Northwestern University.

———. 1969. "Some Spacing Measures of Areal Point Distribution Having the Circular Normal Form." *Geographical Analysis* 1: 15–30.

King, Leslie. 1967. *Statistical Analysis in Geography.* Englewood Cliffs, N.J.: Prentice-Hall.

Neft, David S. 1966. "Statistical Analysis for Area Distributions." *Monograph Series No. 2.* Regional Science Research Institute.

# 3

# SOME PROPERTIES OF BASIC CLASSES OF SPATIAL-DIFFUSION MODELS

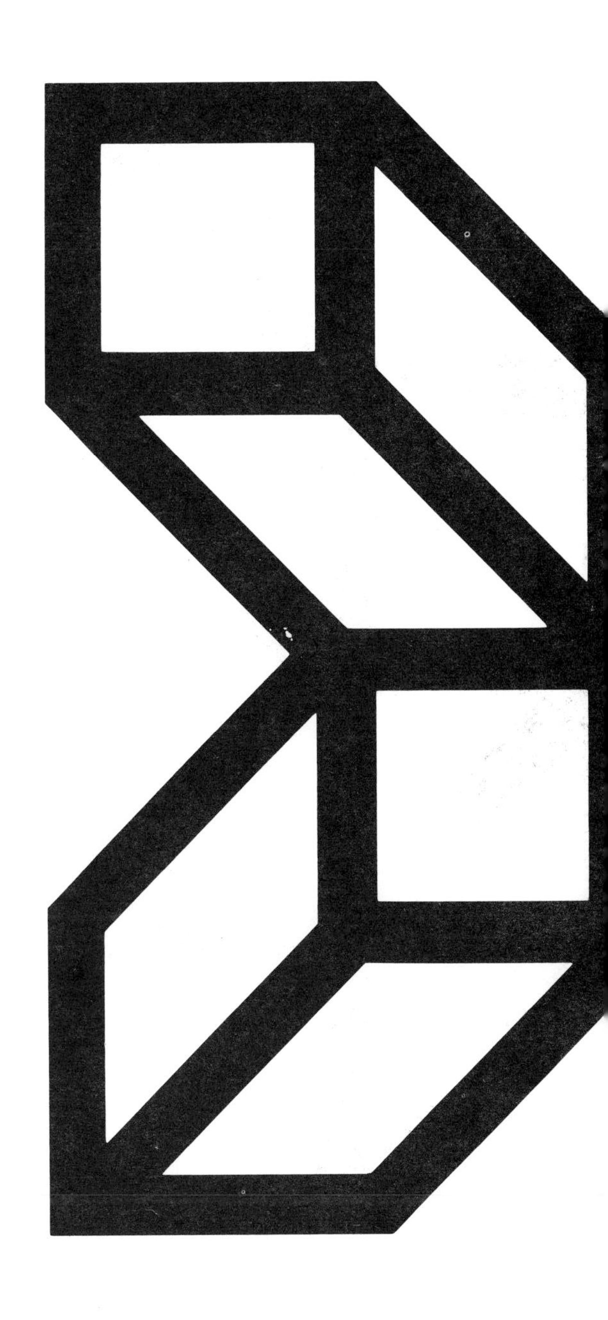

**John C. Hudson**
Northwestern University

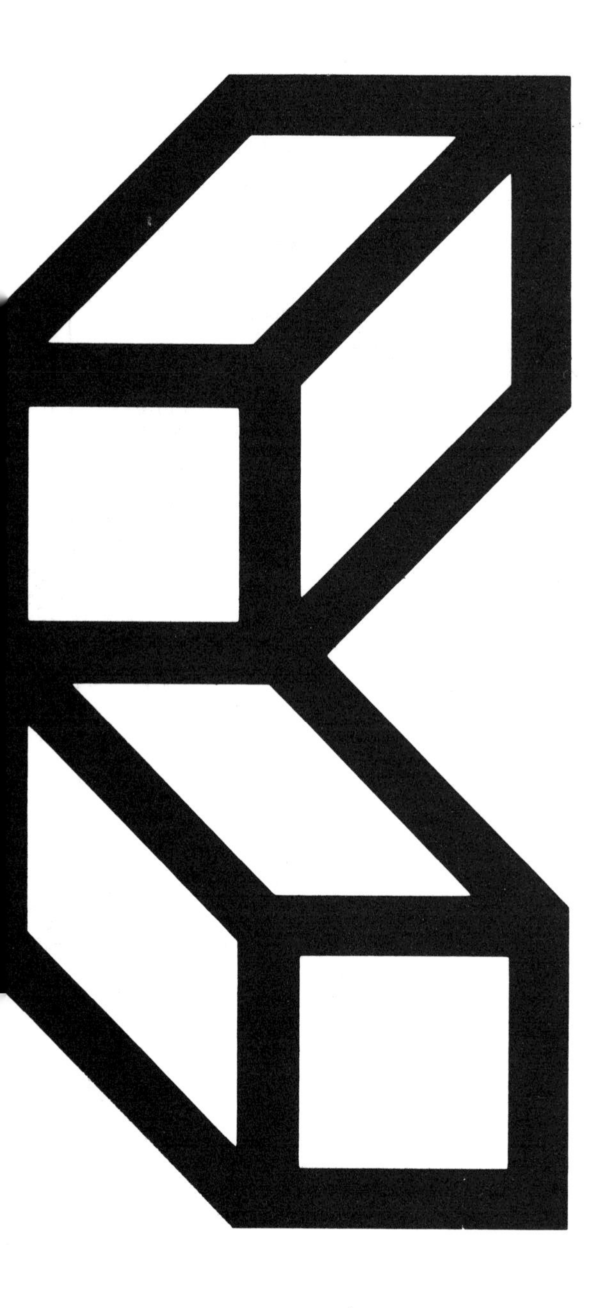

# Some Properties of Basic Classes of Spatial-Diffusion Models

**ABSTRACT**

Spatial-diffusion models may be classified under three broad headings corresponding to particular conceptualizations of communications processes. These are here termed the random-mixing, wave, and hierarchical models. Random-mixing models reflect a society in which interaction is random within certain restraints. An application to Hagerstrand's model of active and passive migration is given. Diffusion-wave models are most closely linked with geographical work. Two basic models are discussed, one postulating a steadily moving wave and the other assuming a wave which decreases in velocity with distance. Means and variances of hearing times for wave models are compared with those for random mixing. The hierarchical models are most suited for the study of diffusion processes in urban systems. Two basic models in this class are considered. They correspond to the two most common types of urban dominance trees. The property of intraorder variability in time until hearing illustrates differences in the two models. It is hoped that future work may be able to rely partly on various properties of the models discussed in this paper.

Most of the models of spatial-diffusion processes that have been proposed to date can be grouped under three headings, each relating to a particular conceptualization of the communications structure of society. One class of models assumes a population which mixes at random and may be divided into various subgroups. Within each of these random mixing takes place, but between them only limited contact exists. These models were developed primarily by Dodd (1956), Rapoport (1963), Coleman (1964), and more recently by Bartholomew (1967). A second group of models focuses on the outward, wave-like spread of diffusion away from a hearth of origin. This wave model has probably had the greatest influence on geographic work, since a similar concept is the basis of the Monte Carlo simulation models. The Hagerstrand-inspired school of diffusion research has focused on this approach. Contributions to methodology in this area have been extensively reviewed by Morrill (1965), Brown and Moore (1969), and Gould (1969). The third category treats diffusion processes as though they take place in a hierarchical structure of communication flows which can usually be graphically portrayed as a semilattice arrangement. Models of diffusion in an urban system are generally portrayed in this manner (Hagerstrand, 1967). This paper is a review of certain aspects of these three classes of diffusion models, in the hope that a systematic treatment of their various properties may serve as a guide for future work. The emphasis is upon the gross, structural properties of the models rather than upon specific interpretations in the context of real processes.

## RANDOM-MIXING MODELS

Regardless of what kind of phenomena is being diffused, one frequently noted characteristic of the temporal behavior of the spread is that the cumulative frequency of knowers, adopters, and so forth, forms an S-shaped or logistic curve. There are a number of arguments why this should be so.[1] A variety of mathematical models have been proposed to explain the process, some of which conflict with each other. Many functions have a cumulative S-shape (most outstandingly, perhaps, the normal distribution), and S-shapedness may be due to a variety of factors. Curve-fitting of diffusion data has yielded little insight into the process since so many processes have this characteristic in common.

Perhaps the simplest formulation leading to the S-curve is due to Rapoport (1963) and to Dodd (1956). Consider a large population of individuals mixing at random. Each time two persons meet they pass a message. Alternatively, consider a population of towns communicating with each other at random, with some process of information transfer taking place. The proportion who are knowers is denoted $p$ and its complement is $q = 1 - p$. Intuitively, $p$ (as well as $q$) depends on time. As-

1. Cassetti (1969), Day (1970), Dodd (1956), Griliches (1957), Rogers (1962), and Tarde (1903) all give different interpretations for the cause of the S-shaped adoption curve.

sume that the rate at which the diffusion is taking place is a constant in time, denoted as k. The probability that a message is passed from a knower to a nonknower in some infinitesimal fraction of time, dt, is the probability that a knower contacts a nonknower, or

$$(1) \quad dp(t)/dt = kp(t)q(t).$$

The solution of the differential equation is

$$(2) \quad p(t) = 1/(1 + ce^{-kt})$$

where $c = q(0)/p(0)$, the ratio of nonknowers to knowers at the start of the process. Since $p(\infty) = 1$, all persons eventually hear. The important assumption is random mixing of the population. Roughly speaking, this model describes interaction in groups with a social or spatial structure in which it is equally likely that any pair of individuals interact. Although it is possible to derive a logistic curve from other assumptions, the models which have described alternative processes have either ignored the nature of the social network, or else depend on the existence of a particular boundary region which limits diffusion in such a way as to produce a leveling-off in the acceptance rate as the innovation spreads over space. On the other hand, not all diffusion-process models lead to a logistic solution.

If the maximum number of adopters is less than the total population, then the relevant equation is

$$(3) \quad dp/dt = kp(a - p)$$

where a is the ceiling of the proportion of adopters. The solution is

$$(4) \quad p(t) = a/(1 + [q(0)/p(0)]e^{-akt}).$$

It is sometimes observed that the potency of the diffusion decreases over time, especially in cases where the item is a fad or something which itself loses popularity. In this case, the constant k in equations (1) and (3) is replaced by a potency parameter, $r(t)$, which itself is a function of time, usually such that $r'(t) < 0$. In this case,

$$(5) \quad dp/dt = r(t)p(1 - p),$$

yielding

$$(6) \quad p(t) = 1/[1 + q(0)/p(0)\exp^{(-\int_0^t r(\tau)d\tau)}].$$

When the ceiling proportion is less than unity, the same modifications are made, *mutatis mutandis,* as in equation (3). $\tau$ is the dummy variable representing integration over the time domain. This curve is termed the harmonic logistic by Dodd (1956). The actual shape of the curve in equation (6) will naturally depend on the function chosen for $r(t)$. Also, the ceiling proportion may be less than unity if $[\int_0^t r(\tau)d\tau]$ converges to some number as t increases. If this is the case, the exponential term remains finite, producing less than total saturation (Rapoport 1963, Sec. 1.4).

### Migration Model

An application of this basic model may be shown in the migration theory of Torsten Hagerstrand.[2] It was Hagerstrand's contention that the migration field about a place was not a static population tributary area, but that it could be thought of as being represented as a chain of connected events. He chose to divide migrants into two categories, one called *active* and the other *passive*. The active migrant seeks a new location because of the prosperity or satisfaction he expects to derive from living in the new location. The passive migrant follows impulses emanating from persons of his acquaintance who have migrated. Hagerstrand based his model on the assumption that all migration from the origin area to some other region, R, is passive, after an initial contact has been made.

Denote, $P_R$ : the population of region R
$V_R$ : the number of vacancies in R between times, $t_i$ and $t_j$.
$I_R$ : the number of residents of R having private contacts with the potential migrants in the origin area, O.

The term vacancy may refer to any type of available, attracting feature in R, the most important of which is a job or a place to live. Hagerstrand's basic formula to predict migration was then,

(7) $M_{t_i,t_j} = k(V_R/P_R)I_R$

or, in words, the number of migrants between time $t_i$ and time $t_j$ is proportional to the number of contacts times the ratio of vacancies to population (vacancy-density) in the receiving area.

One shortcoming of this model is the evident assumption of a limitless supply of population at O. Since the number of potential migrants would itself eventually dwindle, it seems that $I_R$ should depend upon t. Every time a migrant comes from the origin to R, a contact is transferred from being intraregional to interregional.

Let the total population of the origin area (O) which at the beginning is susceptible to moving to R be denoted as a. The model will study the time behavior of the fraction of a that ends up in R. At any time the number of potential migration-producing contacts that involve interplace connections is equivalent to the sum,

(8) [number of susceptibles remaining in O] $\times$ [cumulative number of migrants from O to R] + number of independent contacts $= [a - P_{O,R}(t)][P_{O,R}(t)] + c$

2. See Hagerstrand (1957).

where $P_{O,R}(t)$ is the cumulative number of migrants from O to R at time t.

Assume that a certain fraction of the contacts carry a message encouraging a move from O to R. This fraction will decline with time as the ratio of vacancies to population in R declines, or in other words, as the vacancies become filled. The constant $V_R/P_R$ is replaced by v(t). If no other migration streams to R existed, then v(t) might be proportional to $[a - P_{O,R}(t)]$. If R had more opportunities than could be absorbed by the in-migrants, then v(t) would be a constant function.

The Hagerstrand model may thus be rewritten as

$$(9)\quad dP_{O,R}/dt = v(t)[a - P_{O,R}(t)P_{O,R}(t)]$$

The migration rate to R is equivalent to the (instantaneous) rate of change of R's population due to in-migration from O. The solution is

$$(10)\quad P_{O,R}(t) = a/[1 + c_o \exp(-\int_o^t v(\tau)d\tau)]$$

where $c_o$ is the ratio $[a - P_{O,R}(o)]/P_{O,R}(o)$.

This model may be fit data on the migration of well-defined social groups. When the network of social relations in the home area is tightly knit, the migration process has little disrupting effect and ties with friends and relatives at "home" tend to be maintained. The movement of small religious groups is one example of this type of process.

### Source and Mixing Components

The diffusion process symbolized in equations 1, 3, and 5 consists of two separable components. All individuals are eventually contacted either by the original source of the message or by someone who has directly or indirectly received the message from the source. If there were no diffusion between individuals and interaction only by pairs with the source was possible, then the growth of the process would be described as

$$(11)\quad dp/dt = k(a - p),$$

indicating the probability that a nonknower in a randomly mixing population contacts the fixed source. Integration of equation (11) yields

$$(12)\quad p(t) = a - c_o e^{-kt} \qquad \text{with } c_o = a - p(0),$$

and if the contact rate k is itself a function of time, r(t), then the analogous result is

$$(13)\quad dp/dt = r(t)(a - p)$$

$$(14)\quad p(t) = a - c_o \exp(-\int_o^t r(\tau)d\tau)$$

Equation 12 has no inflection point and rises at a continually decreasing rate. Equation 14 is also nondecreasing, but it may increase at a variable rate.

### Stochastic Model

A stochastic random-mixing model is described in Bartholomew (1967). Some properties of this model are reviewed here for the sake of comparison with the models in the next two sections.

Consider a population of N individuals interacting at random. One person carries a message at the start of the process. At any subsequent time there will be n knowers of the message and $N - n$ nonknowers. A person may become a knower either by contacting the source or by contacting another knower. The probability that the number of knowers increases from n to $n + 1$ in an interval t, $t + \delta t$ is

(15) $P\{n \rightarrow n + 1\} = \lambda_n \delta t, \lambda_n \geq 0.$

As in the deterministic model, the probability of an increase in the number of knowers depends on the number of existing knowers. If the population is interacting with the source at a rate $\alpha$, and mixing together at a rate $\beta$, then

(16) $\lambda_n = (N - n)\alpha + \beta n(N - n) = [\alpha + \beta n][N - n]$

$n = 0, 1, \ldots N - 1.$

The term $\beta n(N - n)$ is similar to equation 3.

The number of knowers at time t is

(17) $P\{n(t) = 0\} = e^{-\lambda_0 t}$

$$P\{n(t) = n\} = \left[\prod_{i=1}^{n-1} \lambda_i\right] \sum_{i=0}^{n} [\exp(-\lambda_i t) / \prod_{\substack{j=0 \\ j \neq i}}^{n} (\lambda_j - \lambda_i)]$$

This distribution is quite complicated to evaluate. Instead, the time until the *n*th person hears, $n = 1, 2, \ldots N$, is studied.

The time interval between the *(i − 1)*th and the *i*th person's receipt of the message is represented as an exponentially distributed random variable, $\tau_i$. Hence, the time until the *n*th person hears is the sum of the *n*, independent and exponentially distributed random variables,

(18) $t_n = \sum_{i=1}^{n} \tau i.$

It can be shown that the average time until the *n*th person hears is

(19) $E(t_n) = \sum_{i=0}^{n-1} [1/(N - i)(\alpha + \beta i)]$

and that for an arbitrary individual this is

(20) $$E(t) = \left[\sum_{n=1}^{N} E(t_n)\right] / N$$
$$= 1/N\{\log_e(N - 1 + \alpha/\beta) - \log_e(\alpha/\beta - 1) + 2\gamma\}$$

where $\gamma$ is approximately 0.5772 (Euler's constant). This term is roughly of the order $(\log_e N)/N$. Therefore, the expected time until hearing decreases as the size of the population gets larger. The larger the group, the more possible sources of the message and the shorter the time until being told.

The assumption of random mixing in the population becomes less plausible as N increases in size due to the restructuring effects of distance and social groups. If a population is partitioned into k subregions within each of which individuals mix at random, and all have equal access to the source, and assuming no contact takes place between the subregions (figure 1), then the approximate effect is to multiply by a factor of k the time required for a fraction of the population (p) to have heard the message.

The deterministic counterpart of equation 16 is found in the usual way. The rate of change of the number of knowers is

$$(21)\quad dn/dt = \alpha[N - n(t)] + \beta n(t)[N - n(t)].$$

Hence

$$(22)\quad n(t) = \{e^{(\alpha+\beta N)t} - 1\}/\{\beta + (\alpha/N)e^{(\alpha+\beta N)t}\}$$

which is a more generalized form of the logistic curve of equation 4.

## DIFFUSION-WAVE MODELS

A large portion of the diffusion research in geography has involved the role of distance as a friction on social communication. Most of this work has been an extension of the work begun by Hagerstrand, especially utilizing his Monte Carlo simulation model.

Consider a population uniformly scattered across a region having density k at every point. If each individual contacts the individuals that are located within some area (dxdy) about himself, then a process results in which a wave of diffusion passes outward across the region from its origin. The assumptions of highly restrictive communication make this a plausible model for diffusion in cultures where communications technology is primitive. This is the kind of model Hagerstrand terms the "innovation wave." It seems to embody the essential ideas of the diffusion process postulated by many anthropologists and cultural geographers.[3] The use of this conceptualization in cultural geography led to the use of terms such as "cultural region," an area within which a certain set of cultural traits has diffused. The topology of the model is simple. There are two regions: an interior one that has been exposed, and an exterior one that has not, or is resistant. *A priori,* the region can have no pockets of nonknowers surrounded by knowers if the environment is homogeneous. When these enclaves are found, some type of resistance or barrier is usually postulated (Carter 1968, pp. 538–63).

3. The diffusionist approach is discussed in Kroeber (1937).

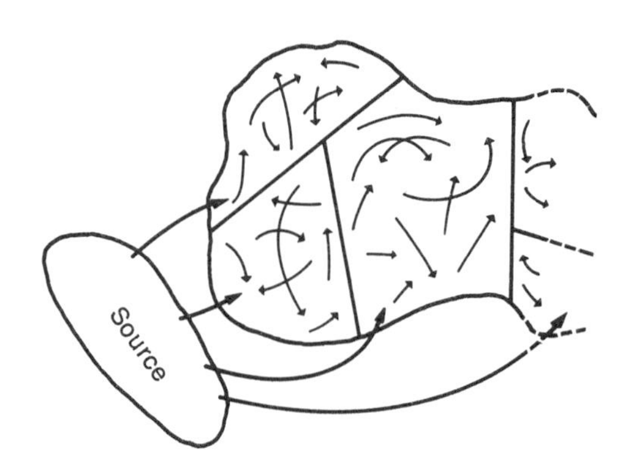


Figure 1. Interaction in a Stratified Population.

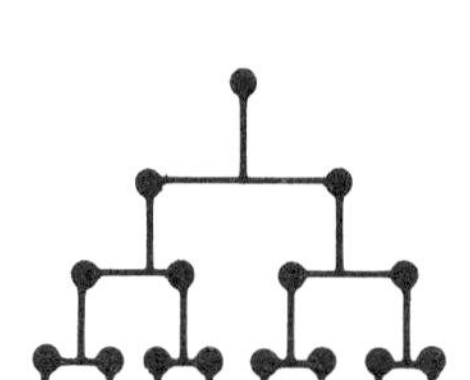

Figure 2. Family-tree distribution. k = 2; m = 4.

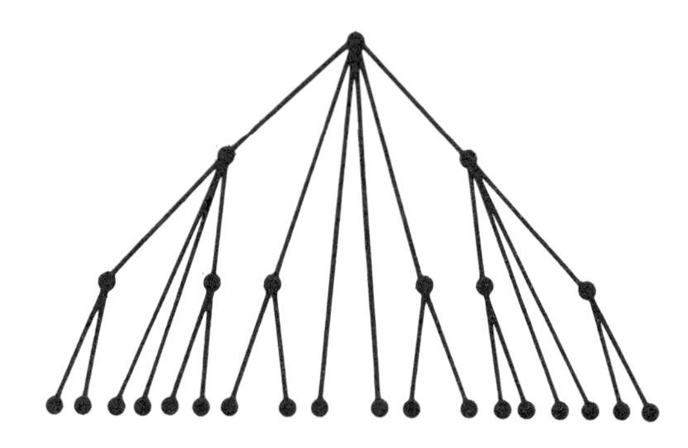

Figure 3. Central-place distribution. q = 3; m = 4.

A rudimentary version of the wave model assumes that the diffusion process involves no resistance, and hence the time of initial contact is the same as the time of adoption, conversion, or whatever.

In a circular region of radius R, with population distributed everywhere at a density k, the number of persons within a radius r of the origin is

(23) $N(r) = k\pi r^2.$

If the diffusion wave advances at a constant rate in time, then

(24) $dr/dt = c.$

By the chain rule for derivatives

(25) $dN/dt = (dN/dr)\,(dr/dt) = 2k\pi rc.$

The time rate of change of the number of adopters is a linear function of distance. In particular, since $dr/dt = c$,

(26) $r(t) = ct + c_o.$

If diffusion starts from an origin point, then $c_o = 0$. Thus by integrating equation 26 with $c_o = 0$,

(27) $N(t) = k\pi(ct)^2.$

The diffusion is completed when the wave reaches R, at time R/c. N is a second-degree function of time (or distance) right up to the limit of population size, with no inflection point.

Rather than having the area be bounded, it is possible that diffusion eventually ceases because the diffusion wave has a decreasing velocity as it moves away from the source. In this case equation 24 may be rewritten as

(28) $dr/dt = a - br$

under the assumption that the decrease in velocity is a linear function of distance from the source. Without introducing any resistance to the innovation, again

(29) $dN/dr = 2k\pi r.$

Solution of equation 28 yields

(30) $r(t) = (a/b)(l - e^{-bt})$

assuming that the initial diffusion begins at a point. The outermost extent of the region is $r = a/b$, which, in practice is never actually attained in a finite time.

Since $N(r) = k\pi r^2$, then

(31) $N(r(t)) = k\pi[a/b(l - e^{-bt})]^2.$

This could also be shown by integrating

(32) $dn/dt = 2(a\sqrt{k\pi N} - bN]$

The eventual number of adopters depends solely upon the ratio between a and b and the population density. The cumulative number of adopters curve is not S-shaped, but is an exponential function similar to, but not exactly like, that of equation 12. The process responsible for equation 28 does not depend on a decreasing rate of spread over time as in equation 5, but on a slowing of the diffusion wave as it reaches farther from its origin. (This, of course, amounts to a decreasing time rate of spread, due to the form of the model.) A likely explanation for this is that, as the innovation moves farther and farther from its hearth of origin, it may become less and less appropriate for adoption, because of gradients in the natural and human environment. Hence, its rate of spread decreases with distance from the hearth. Once again, the communications structure of influence-flows is spatially restricted by assuming contact only between adjacent points.

**Average Waiting Times**

The probability that an individual is arbitrarily within a distance $r_o$ of the origin is

$$(33) \quad F(r_o) = \int_o^{r_o} 2k\pi r dr,$$

since r has the distribution function

$$(34) \quad F(r) = k\pi r^2.$$

The probability that an individual is within the outer limit (R) is

$$(35) \quad F(R) = \int_o^{R} 2k\pi r dr = 1,$$

hence

$$k = 1/\pi R^2.$$

Since

$$(36) \quad dF(r)/dr = 2k\pi r = 2k\pi ct,$$

the expected time until being reached by the wave (waiting time) is

$$(37) \quad E(ct) = c\int_o^{R/c} t(2k\pi t)dt = 2R/3c^2$$

which increases as the radius of the region and decreases as the square of the diffusion-wave velocity. Of course, given r, then $E(t) = ct$, which is completely determined and has no variance. Waiting time is completely determined by distance from the origin. For an arbitrarily chosen individual, the variance in expected time until hearing is

$$(38) \quad V(t) = \int_o^{R/c} t^2(2k\pi t)dt - (2/3R)^2 = R^2/18.$$

For arbitrary c, the expression is

$$(39) \quad V(ct) = c^2\, V(t) = (cR)^2/18.$$

Thus, the variance in time until being reached increases proportionately to the area of the region.

The result in the wave model is exactly the opposite of equation 20 which assumed random mixing. In random mixing, waiting time decreases as population or group size increases. In the wave model, waiting time *increases* proportionately to the square root of area, and hence to the square root of population size.

In the case of equation 28 where the diffusion wave declines in velocity over area, the time that any arbitrarily selected individual is reached is also completely determined by distance and by the factors causing the rate of advance of the innovation to decline with distance. The expected waiting time may be found in the same way as employed above.

Because the total population of the region of radius $R = a/b$ is

(40) $N(a/b) = k\pi(a/b)^2,$

and since eventually all individuals are reached, it follows that

(41) $\int_0^\infty dN(t) = k\pi(a/b)^2,$

which expresses n in terms of area. Since

(42) $r(t) = a(l - e^{-bt})/b,$

(43) $dN[r(t)]/dt = (2k\pi a^2/b)e^{-bt}(1 - e^{-bt}).$

As a check,

(44) $\int_0^\infty dN = k\pi(a/b)^2$

as required. The probability-density function of t then becomes

(45) $n(t) = 2be^{-bt}(1 - e^{-bt}); t \geq 0,$

which is a special case of the rank-variate distribution, sometimes called the Furry-Yule distribution (Feller 1968, p. 450). The first two moments are

(46) $E(t) = 3/2b,$

(47) $V(t) = 3/2b^2.$

The greater the value of b, the more rapidly the diffusion takes place. The parameter (a) does not enter into equation 45, due to the exponential form. In an empirical context, however, it would be the value which related the size of area susceptible to diffusion to the rate of decline of the diffusion wave's advance.

Any application of this model must be in a context reflecting the highly restricted nature of the information-flow process. If this process is one that relies exclusively on face-to-face or very short-distance communication, then the model would be suitable, although still highly simplified. If the information- or influence-flow is not restricted by distance in a wavelike fashion, then the model is inappropriate. These conditions were satisfied in Hagerstrand's early studies of innovation diffusion in rural Sweden but are substantially less plausible in an urban setting.

### Hierarchical Models

Models of diffusion based on characteristics of a hierarchy are much less common than those based on the diffusion wave or upon random mixing. The hierarchical models are most easily expressed in discrete time due to the assumption of a network of discrete nodes. The discussion of the hierarchical models which follows is in terms of a system of urban places, although they may obviously be applied to a wide variety of organizational structures. There are two common ways of characterizing urban dominance structure (figures 2 and 3). For lack of a better term, the first diagram is called the *family-tree distribution.* The second diagram (figure 3) represents the set of influence-flows in one type of Christaller central-place system. This general topological type will be referred to as the *central-place distribution.*

As very crude models of urban influence-flow systems, these two structures represent distinct kinds of dominance trees. The solid lines in the figures indicate the dominance relations, with the most influential center at the top. The nodes represent cities. The *order* of a node refers to its vertical position, with the highest level being the highest order. In central-place theory, each center of a given order has the same population; hence, the levels in the hierarchy can be thought of as a city-size distribution. One city *dominates* another if a chain of dominance relations reaches from one down to the other. A higher-order city *directly dominates* a lower-order one if it is connected by one link.

In the family-tree distribution every city of order j directly dominates a fixed number of cities of order $j - 1$. Every jth order center is linked to the top by the same number of intermediate linkages. In the central-place distribution, every city of order j directly dominates a fixed number of cities of order $j - 1, j - 2, \ldots 2, 1$. This is due to the spatial-economic structure of the central-place model and arises from the postulates of central-place theory rather than from diffusion processes. The family-tree distribution is used as an urban-system model by Beckmann (1968), whereas Christaller (1966), Losch (1967), and Dacey (1965) have treated the central-place model. The city-size distributions that correspond to the two models are, of course, different, owing to their different ratios of higher- to lower-order centers between any two levels.

### Family-Tree Distribution

Suppose a message is passing downwards through an m-order family-tree distribution in which each j-order node directly dominates k nodes of order $j - 1$. In each unit of discrete time the message is passed by the knowers to those nodes which they directly dominate. The number of nodes of each order is given by the geometric series,

$$(48) \quad n(j) = k^j \qquad j = 0, 1, 2, \ldots m - 1.$$

It follows that the number of hearers of the message at any time (t) will be

$$(49) \quad n(t) = k^t \qquad t = 0, 1, 2, \ldots m - 1,$$

due to the deterministic assumption of a lock-step telling order.

The probability that an arbitrarily chosen center hears at time t is

$$(50) \quad f(t) = k^t / \sum_{t=0}^{m-1} k^t.$$

The sum of the geometric series in the denominator is

$$(51) \quad s_m = \sum_{t=0}^{m-1} k^t = (k^m - 1)/k - 1;$$

hence the required probability is

$$(52) \quad f(t) = k^t(k - 1)/(k^m - 1); t = 0, 1, 2, \ldots m - 1.$$

The mean and variance of the time until hearing for an arbitrarily chosen center are, respectively,

$$(53) \quad E(t) = mk^m/(k^m - 1) - k/(k - 1),$$

$$(54) \quad V(t) = [k/(k - 1)^2] - [m^2k^m/(k^m - 1)^2].$$

Both the mean and variance decrease as k gets larger, whereas expected waiting time increases as m increases.

### Central-Place Distribution

The simplest derivation of a telling model for a central-place distribution is that for a Christaller system. In such a system, centers of order j are spaced a distance $q^{(j-1)/2}$ apart, where $j = 1, 2, \ldots m$. Each center then directly dominates $(q - 1)$ of the closest $j - 1$, $j - 2, \ldots 1$ order centers. It is assumed that any center hears the message (first) from only one higher-order center in a discrete time process. This simplifies the problem into a treelike structure. A derivation of the basic model for this system was presented earlier (Hudson 1969). Only the general argument will be presented here.

Due to the city-size and spacing properties of the central-place network, it can be shown that, if telling priorities are based on size-distance considerations, any j-order center is equally likely in the next time interval after it receives a message to pass the message to any $j - 1, j - 2, \ldots 1$ order center that it directly dominates. It is shown that the probability that an arbitrarily chosen place first hears at time t is

$$(55) \quad f(t) = \binom{m-1}{t}(1 - 1/q)^{t}(1/q)^{m-1-t}; \quad t = 0, 1, 2, \ldots m - 1.$$

This is a binomial distribution with mean and variance, respectively,

$$(56) \quad E(t) = (m - 1)(q - 1)/q$$

$$(57) \quad V(t) = (m - 1)(q - 1)/q^2$$

### Intraorder Variability

Equation 55 gives the probability that any arbitrarily chosen center in the urban system will be contacted at time t. This time is completely determined by the number of intervening links between the center and the top of the hierarchy. It is not completely determined by the order of the center; there is a probability distribution of hearing times for an arbitrarily chosen center of a given order. The practical importance of this is that the model states that centers of the same population size differ in the time at which they are reached by the diffusion process. In any urban system with a central-place dominance structure, there is a variability in the message contents held in a given jth order stratum at some point in time. This is shown as follows.

From the previous derivation, the probability that an arbitrarily chosen place first hears at time t is

$$(58) \quad f(t) = \binom{m-1}{t}(1 - 1/q)^{t}(1/q)^{m-1-t}; \quad t = 0, 1, \ldots m - 1.$$

It can be shown that the probability that an arbitrarily chosen place is of order i is given by

$$(59) \quad \begin{aligned} g(i) &= 1/q^{m-1}; \quad i = 1 \\ &= (q^{i-1} - q^{i-2})/q^{m-1}; \quad i = 2, 3, \ldots m - 1. \end{aligned}$$

The joint probability that a place is of order i and first hears at time t is, then,

$$(60) \quad h(t,i) \neq f(t) \cdot g(i),$$

since the time of hearing is not independent of size.

It can be verified that the joint function h(t,i) is

$$(61) \quad \begin{aligned} h(t,i) &= 1/q^{m-1}; \ i = 1; \ t = 0 \\ &= 0; \ i = 2, 3, \ldots m - 1; \ t = 0 \\ &= \binom{i-2}{t-1}(q - 1)^{t}/q^{m-1}; \ t = 1, 2, 3, \ldots i - 2 \\ &= 0; \ t \geq i - 1. \end{aligned}$$

By definition of conditional probability, the marginal density of g given i is

$$(62)\quad h(t|i) = h(t,i)/g(i) = \binom{i-2}{t-1}(1-1/q)^{t-1}(1/q)^{i-t-1}; \quad i \geq 2, t = 1, 2, \ldots i-1,$$

which is itself a binomially distributed random variable with mean

$$(63)\quad E_i(t) = (i-2)(q-1)/q,$$

and variance

$$(64)\quad V_i(t) = (i-2)(q-1)/q^2$$

Variability of hearing times among a set of equal-size places is greatest among the smallest places, and least for the largest places.

In the family-tree distribution, the joint probability that a center is of order i and hears at time t is

$$(65)\quad f(t,i) = f(t);\ t = i-1 \qquad = 0;\ t \neq i-1.$$

The conditional distribution of hearing times, given order is thus

$$(66)\quad f(t|i) = 0;\ i \neq t+1 \qquad = 1;\ i = t+1$$

and is a one-point distribution, with zero variance. The time of hearing is completely determined by order or size of place. The property of order-hearing times illustrates the differences between the family-tree and central-place distributions.

## SUMMARY

In this paper several classes of diffusion models were presented and were discussed in the context of the diffusion processes to which they correspond. Of the three classes of models, those postulating random mixing are the most easily treated in a stochastic framework, as well as deterministically. These models ignore specific channels of influence-flow in the population but may incorporate various interaction rates between and within population subgroups. The classical geographical model of the innovation wave is applicable in the case of a very strong distance friction on the diffusion process. It ignores social structures that are not distance-related. The hierarchical models focus specifically on the organizational aspects of the population and are especially suited to reflect an urban dominance structure in the case of diffusion in a system of cities. All of these models have evident shortcomings. They are too restrictive in their assumptions about the conditions for interaction. They do offer alternative approaches to diffusion problems and are modifiable under a wide variety of conditions. Actual application would have to reflect these modifications. Nevertheless, future work directed toward building a more satisfactory geographical diffusion theory than now exists will probably rely in part on some of the wide variety of properties that are illustrated in these basic models.

## REFERENCES CITED


Bartholomew, David J. 1967. *Stochastic Models for Social Processes.* New York: John Wiley and Sons.

Beckmann, Martin. 1968. *Location Theory.* New York: Random House.

Brown, Lawrence A., and Eric Moore. 1969. "Diffusion Research in Geography: A Perspective." *Progress in Geography* 1:119–58.

Carter, George F. 1968. *Man and the Land.* New York: Holt, Rinehart & Winston.

Cassetti, E. 1969. "Why Do Diffusion Processes Conform to Logistic Trends?" *Geographical Analysis* 1:101–105.

Christaller, Walter. 1966. *Central Places in Southern Germany.* Englewood Cliffs, N.J.: Prentice-Hall.

Coleman, James S. 1964. *Introduction to Mathematical Sociology.* New York: Free Press.

Dacey, Michael F. 1965. "The Geometry of Central Place Theory." *Geografiska Annaler* 47:111–24.

Day, Richard H. 1970. "A Theoretical Note on the Spatial Diffusion of Something New." *Geographical Analysis* 2:68–76.

Dodd, Stuart C. 1956. "Testing Message Diffusion in Harmonic Logistic Curves." *Psychometrika* 21:191–205.

Feller, William. 1968. *An Introduction to Probability Theory and its Applications* Vol. 1. New York: John Wiley and Sons.

Gould, Peter. 1969. "Spatial Diffusion." *Commission on College Geography.* Resource Paper No. 4. Washington, D.C.: Association of American Geographers.

Griliches, Zvi. 1957. "Hybrid Corn: An Exploration in the Economics of Technological Change." *Econometrica* 25:501–22.

Hagerstrand, Torsten. 1957. "Migration and Area." In *Migration in Sweden.* A Symposium, edited by Torsten Hagerstrand and Bruno Odeving. *Lund Studies in Geography* No. 13. Lund, Sweden: C. W. K. Gleerups.

———. 1967. "On the Monte Carlo Simulation of Diffusion." *Quantitative Geography* Part I. Economic and Cultural Topics. Evanston, Ill.: Department of Geography, Northwestern University.

Hudson, John C. 1969. "Diffusion in a Central Place System." *Geographical Analysis* 1:45–58.

Kroeber, A. L. 1937. "Diffusionism." *The Encyclopedia of the Social Sciences* 3:139–42.

Losch, A. 1967. *Economics of Location.* New York: John Wiley and Sons.

Morrill, Richard L. 1965. "Migration and the Spread and Growth of Urban Settlement." *Lund Studies in Geography,* No. 26: Lund, Sweden: C. W. K. Gleerups.

Rapoport, Anatol. 1963. "Mathematical Models of Social Interaction." In Luce, R. D., Galanter, E., and Bush, R., eds., *Handbook of Mathematical Psychology* Vol. 2. New York: John Wiley and Sons. Pp. 495–579.

Rogers, Everett M. 1962. *Diffusion of Innovations.* New York: Free Press.

Tarde, Gabriel. 1903. *The Laws of Imitation.* New York: Holt, Rinehart & Winston.

# 4

# ON RURAL SETTLEMENT IN ISRAEL AND MODEL STRATEGY

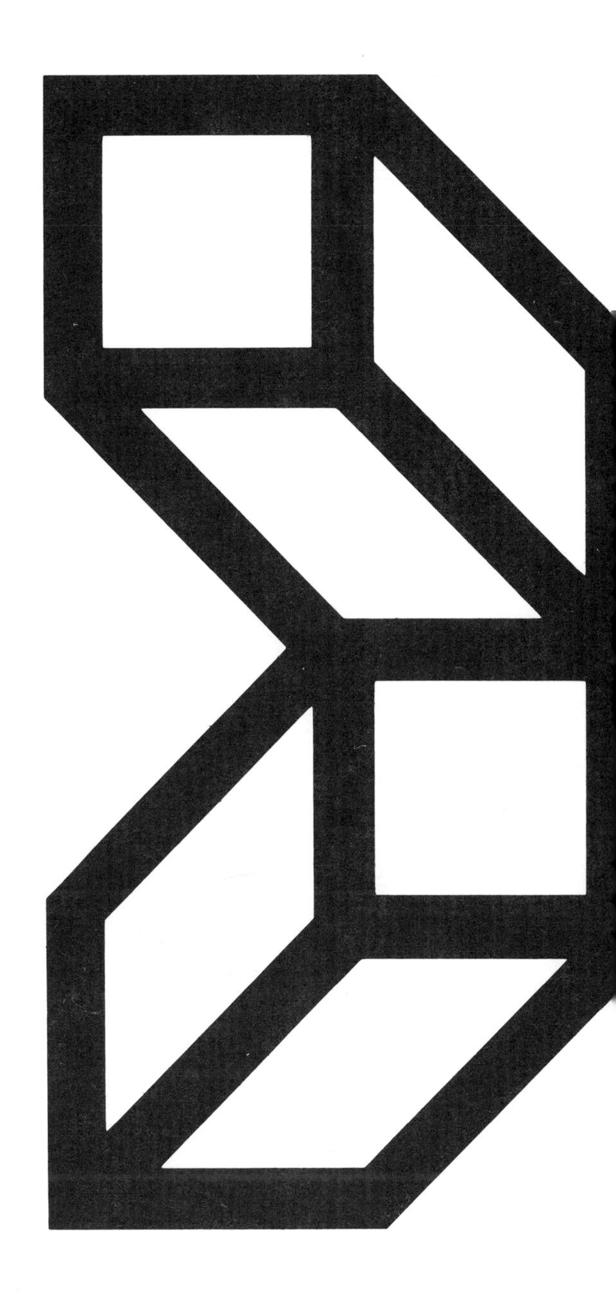

**Lawrence A. Brown**
The Ohio State University

**Melvin Albaum**
University of Colorado

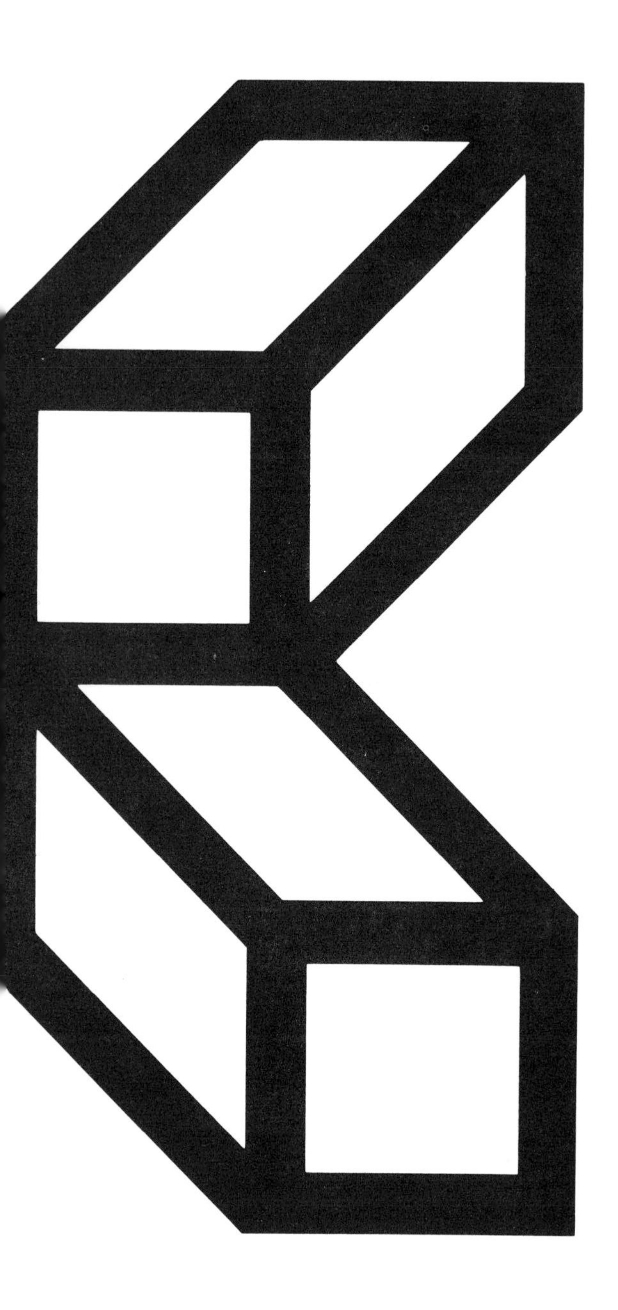

# On Rural Settlement in Israel and Model Strategy

## ABSTRACT

This paper is concerned with the study of growth processes in a spatial context. The particular situation examined is that of the development of rural settlement patterns in Israel. Drawing upon empirical evidence of that development, a model is presented that appears capable of replicating the events of the real world and serving as a medium for evaluating alternate strategies of development. Of particular importance, it seems, is the identification of the interrelationships between changes in settlement pattern and changes in environment, a significant component of which is the system of settlements, and the effects of individual and government behavior on those relationships. The model is presented in a heuristic framework, and no attempt at direct application is made.

This paper presents a Monte Carlo simulation model designed to replicate the spatial patterns of rural settlement in Israel, as they have occurred through time, and to study the interrelationships which exist between such patterns and regional and national growth processes.[1] The relevance of this approach has been suggested by Boulding, in a general discussion of growth process: "Growth creates form, but form limits growth. This mutuality of relationship is the most essential key to the understanding of structural growth."[2] If one considers settlement in a spatial context, we find evidence of this paradox, for environment influences or limits the location of settlement, but settlement in turn alters environment; this, in turn, affects location in the next time period, and so forth.

As geographers, we seek knowledge of the dimensions of the limiting paradox posed by Boulding as it relates to location or locational patterns on the earth's surface. Such an approach should provide insight into strategies for effective division or use of available land area, given the changing confines of the environment. The latter gain, of course, implicitly points toward planned and controlled growth programs. In this light, rural settlement in Israel is an especially appropriate case since it has involved a high degree of central planning and policy implementation. Furthermore, the study of an area where planned development has taken place is more likely to result in identification of the relationships between act, situation, and the limiting aspects of growth.

A Monte Carlo simulation model has great utility for such an examination: once the relevant processes have been embodied within the model, as judged by its ability to replicate observed patterns, the model may be modified at will by the researcher; thus he may manipulate any element of the process in order to examine its effect upon other elements and upon final patterns. In presenting the model, the intention of this paper is to provide a framework for the study of growth processes in spatial situations, as well as to outline a suitable model. These topics should have significant heuristic value to the student of locational processes in a growth and development context.

## BACKGROUND: ISRAELI RURAL SETTLEMENT

There are basically four types of Jewish rural settlements: *moshava* (private small-holder's settlement), *kibbutz* (collective settlement), *moshav* (cooperative settlement),[3] and central place.[4] The earliest of these, the *moshava,* was the only form of settlement between 1878 and 1909, with the last one established in 1920. *Moshavot* are composed of privately owned farms.[5] The economy of the *moshava* is labor intensive, with little capital ex-

1. Several aspects of this paper were derived from work by Sven Herner, Department of Geography, The Royal University of Lund, Sweden. A number of helpful comments were provided by Yehoshua Cohen, Department of Geography, University of Chicago. Responsibility of statements made in this paper, of course, lies solely with the authors.

2. K. E. Boulding, "Toward a General Theory of Growth," *Yearbook for the Advancement of General Systems Theory,* Vol. 1, 1956, p. 72.

3. The plural of *moshava* is *moshavot;* the plural of *kibbutz* is *kibbutzim;* the plural of *moshav* is *moshavim.*

4. There are many published works pertaining to Israeli rural settlement. See, for example, J. Ben-David, ed., *Agricultural Planning and Village Community in Israel* (Paris: UNESCO, 1964); H. Darin-Drabkin, *Patterns of Cooperative Agriculture in Israel* (Tel Aviv: Israel Institute for Books, 1962); E. Orni and E. Efrat, *Geography of Israel* (New York: Daniel Davey, 1964), especially pp. 199–216. For a summary of the basic distinguishing characteristics of Israeli rural settlement types, see H. Halperin, *Changing Patterns in Israel Agriculture* (London: Routledge and Kegan Paul, 1957), chapter 7, and D. Weintraub, M. Lissak, and Y. Azmon, *Mashava, Kibbutz, and Moshav: Patterns of Jewish Rural Settlement in Palestine* (Ithaca: Cornell University Press, 1969).

5. *Moshavot* and a very small number of *moshavim* adhere to the principle of private ownership of the land. All *kibbutzim* and the vast majority of *moshavim* are established on lands owned by the Jewish National Fund (a public institution) and the government. These lands are managed by the Israel Land Authority, founded in 1958. These lands are leased to permanent settlements for periods of forty-nine years, charging rents which are far less than the average economic rent. The land rent policy of the ILA is thus the dominant system in Israel. For further discussion on this subject, see S. N. Eisenstadt, *Israeli Society* (New York: Basic Books, 1967), pp. 78–80; and A. Bonné, "Major Aspects of Land Tenure and Rural Social Structure in Israel," in *Land Tenure,* ed. K. H. Parsons, R. J. Penn, and P. M. Raup (Madison: University of Wisconsin Press, 1956), pp. 111–16.

penditure. The form of these villages is quite similar to the agricultural villages found in middle Europe and probably reflects the origins of a majority of the early settlers.

*Moshava* settlers tended to select areas of relatively high fertility which were near existing Arab or Jewish market places. The density of the Arab population was also a factor in site selection. However, many factors which were important to location decisions of later settlements—for example, the improvement potential of the area and Arab tenacity or resistance with regard to the land—played no role in determining the locations of the *moshavot.* Prospects of soil improvement were quite limited owing to the lack of appropriate technology and capital, and Arab tenacity with respect to the land was of little importance because Jewish settlers were few and paid well for the land. During this initial period of Jewish settlement in Palestine, there was no popular ideology of Israel as an existing nation, and the Arabs felt little threat or competition for the land. Arab hostility, therefore, which was to become an increasingly important factor in later Jewish colonization efforts, was relatively nonexistent.

The *kibbutz* first appeared in 1909, but it predominated Jewish settlement during the 1930s and early 1940s.[6] The *kibbutz* is associated with the rise of the ideology of Israel as an independent political entity. As such, it performed certain economic and strategic functions. One of these involved the creation of an infrastructure upon which an integrated set of new settlements could be based. A second entailed the occupation of what was conceived of as the desired national territory, especially border areas. Associated with this second task was the desire to establish settlements in hostile Arab regions with high strategic or resource value. Economically and socially, the *kibbutz,* with its somewhat isolated position, became an independent entity with few links to other than major centers, such as Tel Aviv, Haifa, and Jerusalem and, where feasible, to neighboring *kibbutzim.* Production and consumption are collective, with the former characterized by heavy inputs of capital and relatively small use of labor. Moreover, industrial as well as agricultural production has become important in the more recent period.[7] The goods produced are either to support the *kibbutz* population or are of a type that can bear the high transportation cost to larger centers. Production is consequently highly specialized and concentrated in export goods, and the service sector is highly developed. To a great extent, the *kibbutz* type of settlement performs many tasks that are normally found in high-order central places.

As a consequence of its form and function, the locational criteria for the *kibbutz* are very different from those for the *moshava.* Natural fertility is of little importance,

6. For an interesting discussion of the historical evolution of these collective settlements, see E. Orni and E. Efrat, *Geography of Israel,* and Weintraub, Lissak, and Azmon, op. cit.

7. H. Halperin, *Changing Patterns,* pp. 110–11.

whereas improvement potential is of much greater concern since one task of the *kibbutz* is to pave the way for future settlement. Nearness to a market is also of little importance to the *kibbutz,* but, for reasons of national strategy, nearness to the desired national boundary, access to natural resources, or proximity to places of high Arab population density are locational criteria. Arab tenacity might also be looked upon as a locational advantage for reasons of national strategy, but more often it was so strong as to pose a disadvantage. One other important criterion, for both economic and strategic considerations, is the nearness of other *kibbutzim.*

The *moshav* was the dominant Jewish settlement type during the late 1940s and the 1950s, although the first one was founded as early as 1921. Unlike the *moshava,* which is completely private, and the *kibbutz,* which is completely collective in economic and social structure, the *moshav* is based upon cooperative principles.[8] The function of the *moshav* is primarily agricultural, and most of the marketing and purchasing are performed on a cooperative basis. Produce is marketed mainly in cities and in nearby towns, in addition to foreign export. Although some produce does find its market at *kibbutzim,* this is a relatively minor market. Land is owned privately in some *moshavim,* and jointly leased from the Israel Land Authority in most.[9] The total land area of the settlement is generally larger than the *moshava,* but smaller than the *kibbutz.* As in the *moshava,* labor usage tends to be intensive, but the use of capital and machinery is much greater, although less than is found in the *kibbutz.*

Since the *moshav* is basically an agricultural unit, natural fertility is an important locational criterion. However, by the time the *moshav* became the predominant type of Jewish settlement, most areas having good soils were already occupied. The remaining areas with good soil were in a state of serious neglect. Since great expansion of Jewish agriculture was planned, improvement potential was in reality a more important locational criterion than natural fertility. Also, if other factors were favorable, *moshavim* tended to locate near *kibbutzim,* which provided protection and often a supply center. This symbiotic relationship occurred quite frequently, since *kibbutzim* tended to locate on or near land with greatest improvement potential, as noted above. In settled areas, however, *moshavot* or central places also provided an attraction for *moshavim.* On the other hand, a high Arab population density in combination with a high Arab tenacity were negative locational factors.

Since the mid-1950s, new rural Jewish settlements have been primarily of the central place type, integrated with new *moshavim,* within a framework of regional development programs. Generally, the rise of central place type settlements in rural Israel is a totally planned phenomenon.[10] These settlements serve at least one of sev-

8. For a brief discussion of the basic principles of these cooperatives, see M. Albaum, "Cooperative Agricultural Settlement in Egypt and Israel," *Land Economics,* Vol. 42 (1966): 221–25.

9. See footnote 5.

10. See, H. Darin-Drabkin, op. cit., chapter 8. For a general discussion of agricultural and rural planning in Israel, see, R. Weitz, *Agriculture and Rural Development in Israel: Projection and Planning* (Rehovot: National and University Institute of Agriculture, 1962).

eral purposes: (1) to service the agricultural areas; (2) to develop natural resources by providing residence and service sites for labor; (3) to increase the population of remote areas for purposes of national strategy; (4) to control expansion of the large cities; and (5) to integrate all parts of Israel into a functional whole. Consequently, several situations provide positive locational factors for central places. One such situation is the existence of a *moshav* or *moshava* area with no Jewish central place or *kibbutz.* However, central places might also be located in a *kibbutz* area to satisfy the purpose of increasing the population in remote areas, even though the *kibbutz* itself might already provide many traditional central-place functions. Also, since the *kibbutz* members frequently are not willing to cooperate in integration efforts which threaten to alter the basic form and philosophy of the *kibbutz,* a central place may be located nearby to provide the interarea services necessary for an integrated region. The presence of natural resources which have not been developed because of a lack of labor in the surrounding areas also provides a positive locational factor for central places; so too does an area which is suitable for *moshavim* but which has not been developed. In the latter case, the central place will pioneer an area's development rather than be superimposed upon an existing settlement structure.[11] Distance to other central places, of course, also must be considered, as well as the distance from the national border. With regard to the latter factor, there is a tendency to integrate those areas closest to the border first, largely for reasons of military strategy.

It is apparent from the foregoing discussion that settlement location was, and still remains, an important aspect of activities related to establishing and maintaining the state of Israel; consequently, the type of settlement utilized at any particular time and, more important, the location of the settlement are very much related to the objectives of national policy. Generally, four stages of this policy may be identified. The first stage is really one of no policy. From 1878 through 1909, during which time the *moshavot* came into being, the notion of Israel as a nation-state was not publicly acknowledged and, for the most part, settlement was uncontrolled. There continued to be some uncontrolled settlement of the *moshava* type until 1920, but after 1909 and until 1925, the second policy stage dominated. This stage favored the *kibbutz.* It was characterized by the rise of Zionism and a Marxist-type socialist ideology, with financial support often coming from European Jewish communities. Another characteristic of this period was the increasing hostility and tenacity of the Arabs. This element became extremely important during the third policy stage, which dates from 1925 until 1945. This period is characterized by the founding of both *moshavim* and *kibbutzim,* but

11. In the Arab region, for example, because of the lack of ample water supply to support agricultural settlements in the initial stage, the decision was made to establish the regional town as the first stage in the settlement of the region. Agricultural settlements were then to be established as additional water pipelines and deep wells became operational.

Table 1. *Policy Stages of Israeli Settlement*

| Policy Stage | Dates | Settlement Type |
|---|---|---|
| 1 | 1878–1909 (–1920) | *moshava* |
| 2 | 1909–1925 | *kibbutz* |
| 3 | 1925–1945 | *moshav* and *kibbutz* |
| 4 | 1945–present | *moshav* and central place |

the latter were dominant, both because of the importance of ideological motivation among the settlers and because of frequent friction with the local Arab populace. Evidently, one cause of this friction was a more aggressive and nationalistic policy on the part of the Jewish community, which now had serious intentions of founding a state of Israel. Although some *kibbutzim* were established after 1945, primary efforts turned toward filling in the skeletal structure created by the *kibbutzim* and integrating the nation into some functional whole, especially in the southern portions of Israel. As a consequence, the *moshav* and, later, the central place became the dominant forms of settlement.

Some important facts of each policy stage are presented in table 1.

Of all settlements, excluding central places, 10 per cent are *moshavot* in stage one; 7 per cent are *kibbutzim* in stage two; 28 per cent are *kibbutzim* and 10 per cent are *moshavim* in stage three; and 45 per cent are *moshavim* in stage four. Thus, the *moshav* is the most common type of settlement, and the third and fourth policy stages are the most important settlement eras. An indication of the spatial patterns characterizing all settlements for each of the four policy stages is to be found in figure 1.

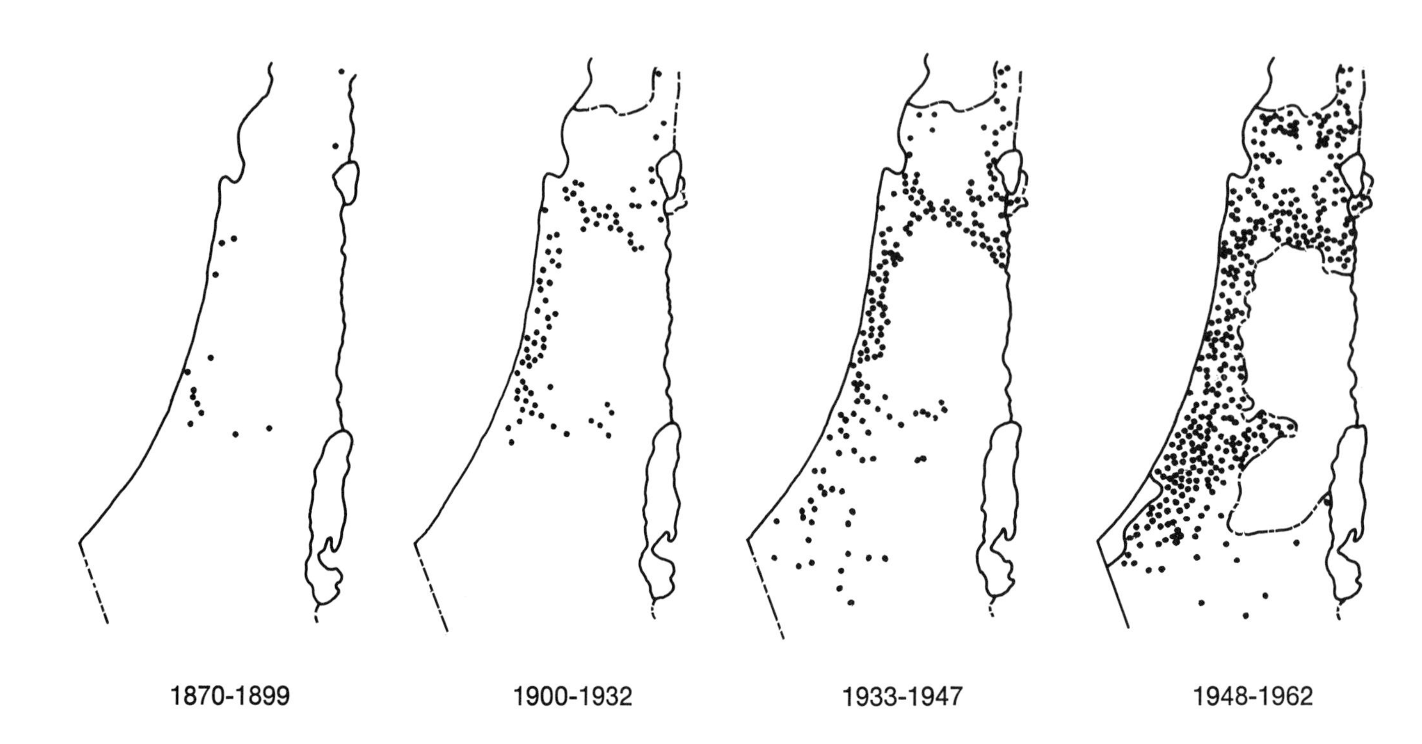


Figure 1. Stages in the Spreading of Jewish Settlement over the Country. Adapted from E. Orni and E. Efrat, *Geography of Israel* (New York: Daniel Davey, 1964).

## A SIMULATION MODEL FOR RURAL SETTLEMENT IN ISRAEL

The principles, processes, and patterns summarized above, are now represented in the form of a Monte Carlo simulation model. The major operational steps of the model serve as focal points for the presentation (figure 2). For convenience, *moshavot* will be symbolized as mt, *kibbutzim* as k, *moshavim* as m, and central places as cp.

1. *Determine the type of settlement to be located during time t.* Since the type of settlement is mostly a reflection of government policy, or lack of it in the case of the *moshavot,* this stage will be exogenous to the model. Accordingly, if each t represents one year and $t = 1 =$ 1878,

$t = 1 \ldots 31 = mt$
$t = 32 \ldots 43 = mt \cdot k$
$t = 44 \ldots 49 = k$
$t = 50 \ldots 70 = k \cdot m$
$t = 71 \ldots 91 = m \cdot cp$

The model might be refined so as to determine policy, and hence type of settlement, according to the actual conditions existing each year. Some of these conditions are exogenous to the model (for example, the political stance of the British parliament); some would be dependent upon the model's predictions of locational decisions for previous time periods (Arab tenacity regarding land, for instance, which is a factor in determining settlement type, may vary according to the amount of Jewish settlement). This refinement would permit examination of the effects of the larger political and social environment within which the location of settlements is decided. Another refinement is to fit the probability density function for each type of settlement as it varies over time and then to determine the settlement type by a Monte Carlo sampling scheme. This would recognize the fact that settlement type eras are not mutually exclusive.

2. *Determine the number of new settlements of the type determined in (1).* A very simple model is to designate the number of new settlements as an exogenous variable. This might be justified by the argument that we are primarily interested in spatial patterns; but, since the temporal pattern has spatial implications in this case, a more complex model, which computes the number of new settlements as a function of other variables, is desirable. The computational procedure will vary according to the type of settlement, but in all cases the first step is to determine the maximum number of settlements of each type which may come into existence during each time period. If we designate

$I_{t-1}$ as the immigration to Israel during time $t - 1$

$E_{t-1}$ as the excess population in established settlements at time $t - 1$

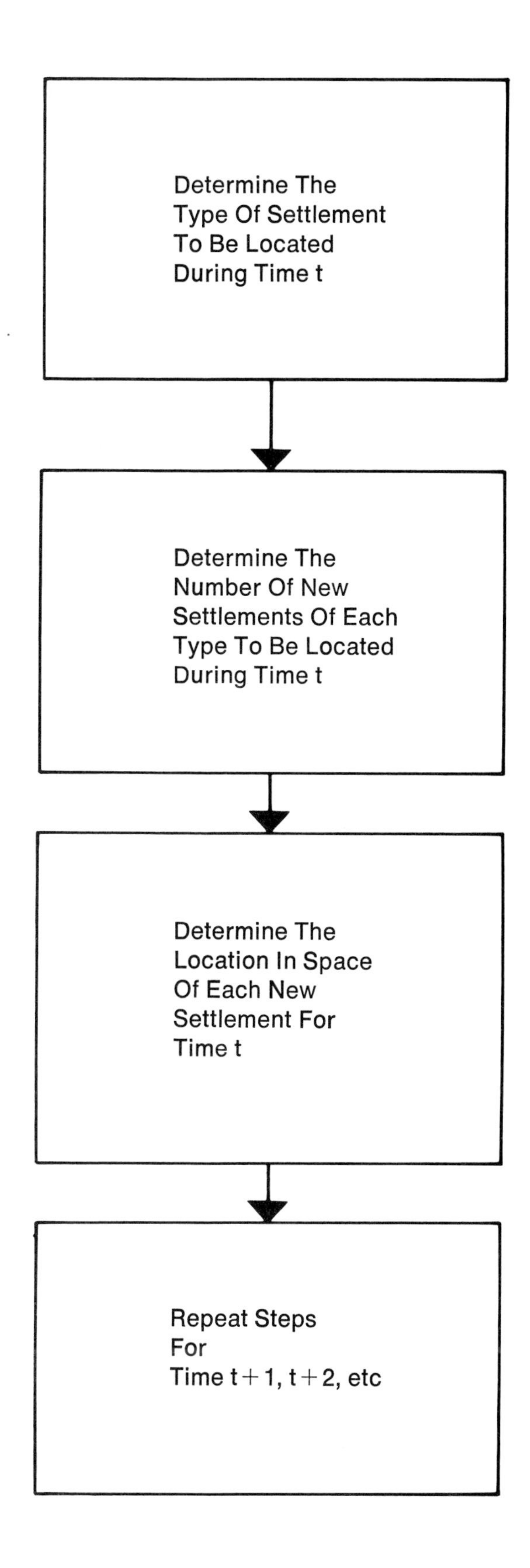


Figure 2. A Flow Diagram of the Major Operational Steps of the Simulation Model for Rural Settlement in Israel.

$\beta$ as a constant which indicates the proportion of $I_{t-1}$ who settle in the regions treated by the model

$\alpha$ as a constant which indicates the proportion of $E_{t-1}$ who consider resettling in the regions treated by the model

$\phi_j$ as the minimum number of people necessary for each type of settlement, where $j = m, k, mt$, or $cp$

then, MAX $j_t$, the maximum number of settlements of type j in time t, is

$$\text{MAX } mt_t = \frac{\beta I_{t-1} + \alpha E_{t-1}}{\phi mt}$$

$$\text{MAX } k_t = \frac{\beta I_{t-1} + \alpha E_{t-1}}{\phi k}$$

$$\text{MAX } m_t = \frac{\beta I_{t-1} + \alpha E_{t-1}}{\phi m}$$

$$\text{MAX } cp_t = \frac{\beta I_{t-1} + \alpha E_{t-1}}{\phi cp}$$

Thus, the maximum number of settlements is always a function of the total number of people who might inhabit those settlements and the minimum population requirement of each settlement type.

The next task is to reduce the maximum number of settlements, since that number will occur only under optimum conditions. This is done by considering the allocation of funds by the government and semigovernment agencies, to the settlement program.

Thus,

$$_tN_{mt} = \text{MAX } mt_t - bA_t$$
$$_tN_k = \text{MAX } k_t - bA_t$$
$$_tN_m = \text{MAX } m_t - bA_t$$
$$_tN_{cp} = \text{MAX } cp_t - bA_t$$

where

$_tN_j$ is the actual number of settlements of type j during time t,

$A_t$ is the time t allocation of funds to the settlement program

b is a constant to be determined from the empirical data

and a linear function is assumed. Accordingly, we hypothesize that the actual number of each type of settlement is a direct function of the "population pressure" as restricted by the availability of funds to relieve this pressure.[12] Presumably, if insufficient funds were available, the population would be absorbed by the larger cities.

The time periods which are characterized by the occurrence of two types of settlement (time periods 32

12. The model assumes that the actual number of each type of settlement is restricted by both capital inputs and labor inputs. It might be useful, however, to distinguish types of inputs, some of which were more critical than others. The supply of skilled agriculturalists and farm managers is one example.

through 43, for example) present some problems with regard to the model presented thus far. It is necessary to incorporate a device which will allocate the total number of settlements between two different types of settlement. Three suggestions are offered. The first is to determine the number of each type according to a proportion given from empirical data. The second is to determine the allocation according to the existing settlement pattern, the social and physical environment, and the policy goals of the governing body. The third suggestion is to determine the allocation according to the nationality characteristics of the immigrants and to the type of settlement from which came the excess population. Of course, the device or model component which would most closely replicate the actual process would include both suggestions two and three.

As a further refinement for the part of the model which determines the number of new settlements, one might consider $\alpha$ and $\beta$ as varying over time and as being functions of factors, such as job opportunities or land availability in large cities. $\alpha$ and $\beta$, it will be recalled, are constants which indicate, respectively, the proportion of the excess population in established settlements who consider resettling in the regions treated by the model, and the proportion of the immigrants to Israel who settle in regions treated by the model. The total number of immigrants might also be a function of external conditions, such as British policy toward Jewish immigration.

3. *Determine the location in space of each new settlement for time t.* The location of each new settlement is seen to depend upon two sets of factors. The first set consists of factors which are independent of the pattern of Jewish settlement. These are called *fixed attributes,* because their value for any parcel of land *is not* affected by changes in the pattern of rural Jewish settlement. Thus, the elements of this set are either static, such as natural fertility, or they are affected primarily by conditions that the model does not simulate, Arab population density, for example.

The second set of factors are directly related to the pattern of Jewish settlement and are termed *associational attributes.* Their value for any parcel of land is affected by changes in the pattern of rural Jewish settlement. One obvious variable of this set is the proportion of available land in each place i. Another important variable in this set is the magnetism of a type of settlement in time $t - 1$ toward the same or a different type of settlement in time t. These relationships, which are detailed in the empirical description section, may be summed up as a connection-type matrix, where 1 indicates that the presence of one settlement type (the row) will induce or attract another settlement type (the column) to locate nearby. Conversely, O indicates no such attraction. Thus we have the matrix:

| Prior Settlement \ New Settlement | mt | k | m | cp |
|---|---|---|---|---|
| mt | 1 | 1 | 1 | 0 |
| k | 0 | 1 | 1 | 1 |
| m | 0 | 0 | 1 | 1 |
| cp | 0 | 0 | 1 | 1 |

In practice, both sets will be treated together to form a single probability of receiving a settlement of each type for each space for each time period. However, it is necessary to treat each type of settlement separately, since different factors are important for each. If we define

$Z_{it}$ as the proportion of available land in place i which is already settled in time t

$F_i$ as the natural fertility of place i

$IP_{it}$ as the improvement potential of place i in time t

$D_{it}$ as the distance of place i from the desired national boundary or the perceived limit of settlement at time t

$A_{it}$ as the Arab settlement factor of place i in time t, reflecting both its Arab population density and its number of Arab settlements, assuming therefore that the two are closely related

$T_{it}$ as the tenacity of the Arabs in place i in time t, a factor which may be collapsible into $A_{it}$

${}_{j}N_{it}$ as the effective number of Jewish settlements of type j in place i by time t, where j = mt, k, m, or cp, and where the term "effective number" gives recognition to the fact that we consider settlements in cells other than i which affect the suitability of new settlement in place i

$L_{it}$ as the labor shortage in place i in time t, which will be primarily related to the natural resource store of place i

Then, from the empirical description section of this paper, the following functional relationships hold, where ${}_{j}S_{it}$ is the probability that a settlement of type j appears in place i during time t:

$${}_{mt}S_{it} = f(Z_{it}, F_i, A_{it}, {}_{mt}N_{it})$$

$${}_{k}S_{it} = f(Z_{it}, F_i, IP_{it}, D_{it}, A_{it}, T_{it}, {}_{k}N_{it})$$

$${}_{m}S_{it} = f(Z_{it}, F_i, IP_{it}, A_{it}, T_{it}, {}_{k}N_{it}, {}_{m}N_{it}, {}_{cp}N_{it})$$

$${}_{cp}S_{it} = f(Z_{it}, D_{it}, A_{it}, {}_{m}N_{it}, {}_{cp}N_{it}, L_{it})$$

The estimates of each of these variables, of course, will present some problems in themselves. For example, what is the exact relationship between distance separating place i from a *kibbutz* and the attraction of place i for a new *kibbutz*? These problems must be worked out from empirical data. Also, the problem of combining many individual variables to produce a single probability ${}_{j}S_{it}$ must be considered. Exactly what will the functional relationship be? The best approach would seem to be the use of regression techniques, where the b values indicate the weighting of each variable.

## UTILIZATION OF THE MODEL

By taking an overview of the model as it has been presented, four general types of factors which influence emerging settlement patterns may be identified:

(1) Type of settlement (e.g., *kibbutzim*) and number of settlers.

(2) Budgetary allocation for the settlement program.

(3) Associational attributes of each place (as defined in step 3), e.g., the type of settlement already located in or near each place.

(4) Fixed attributes of each place (as defined in step 3), e.g., soil fertility, availability of water.

A fifth factor (5), which is not explicit in the model, is the planner's evaluation of the fixed and associational attributes of each place and their relevance in his decision concerning which type of settlement to be placed in any particular location.

These general types of factors can be classified as either being unaffected by the settlement pattern (that is, *exogenous* to the system of settlements)—which include 1, 2, and 5—or being affected by the settlement pattern (that is, *endogenous* to the system of settlements); they include 3 and 4. The utility of these generalizations is that they simplify the decision about which elements of the system may profitably be examined as levers to be used by the planner in molding the system as he wishes. It is clear, for example, that fixed attributes of each place (4) do not present any leverage to the planner, since, like resources, he can do no more than use them prudently. Type of settlement and number of settlers (1) also are factors which are probably beyond the planner's control, since they are most strongly determined from external political conditions. However, associational attributes of each place (3) do provide the planner with leverage because they are directly affected (altered) by his actions, and the degree to which they limit growth is likely to be a function of the way in which they are used. The planner should consequently attend to the effect of his decisions upon associational attributes. Other elements of the system that can be more or less directly controlled by the planner are the budget allocations for the settlement program (2) and, his evaluation of the fixed and associational attributes of each place and their relevance in his decision of which type of settlement to place at a given location (5). One might even argue that these are the most basic factors. Consequently, in manipulating or changing elements of the model in order to examine their effects upon the spatial distribution of settlement, this area of the system should receive special attention.

The best criterion for evaluating such changes is the degree to which significant alteration occurs in the final

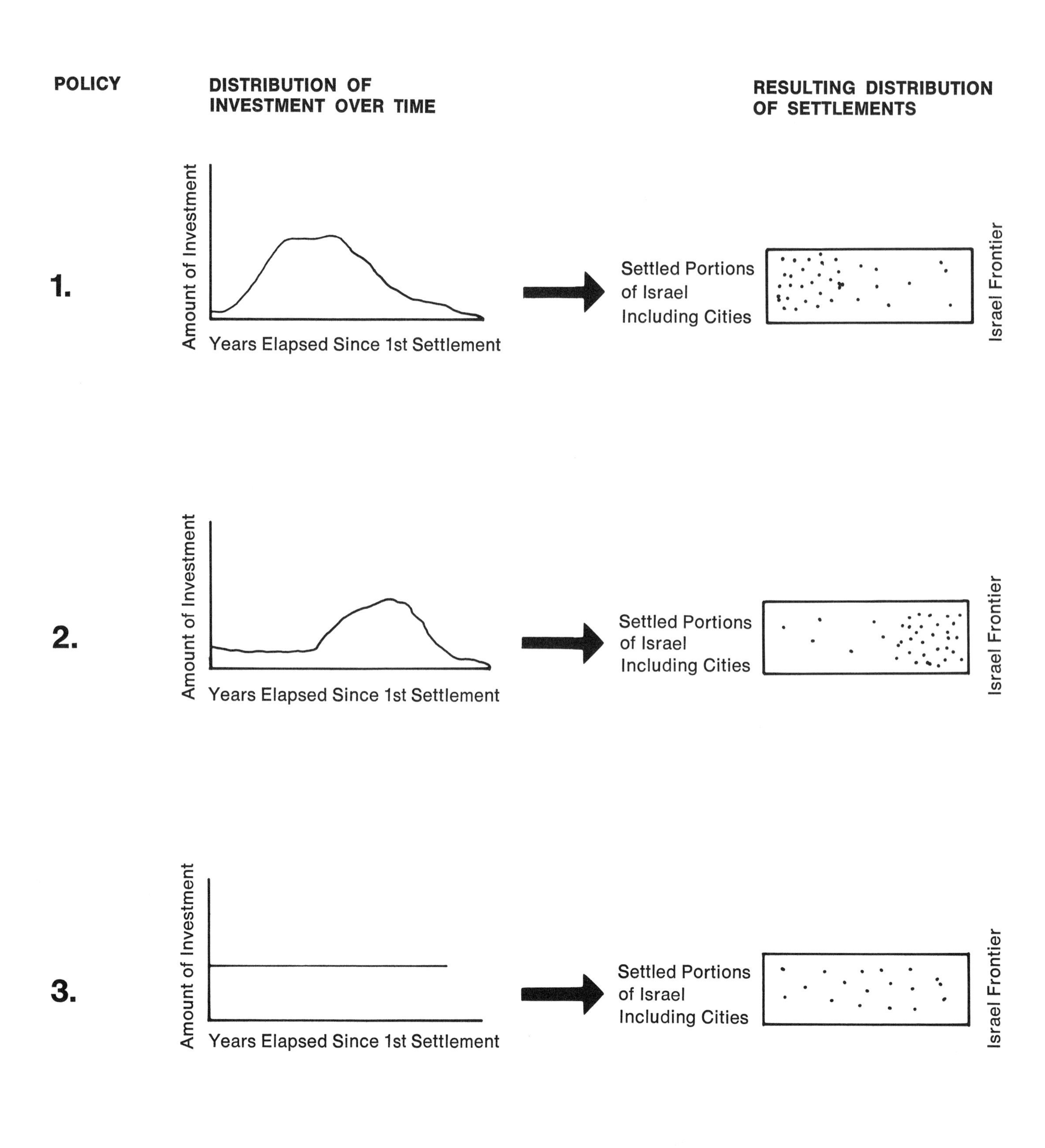


Figure 3. The Spatial Effect of Different Investment Policies, Hypothetical Example.

pattern. This may be otherwise stated as the degree to which the system is unstable vis-à-vis the variable being manipulated (perturbated). For example, if fixed attributes of each place have little effect on the ultimate pattern of settlements, manipulating the planner's evaluation of fixed attributes will also have little effect on the ultimate pattern and therefore need not be considered. In most cases, however, the decision about whether the settlement pattern has been significantly altered (that is, whether the system is unstable vis-à-vis the variable being manipulated) will vary according to the type of outcome considered optimal, and this, in turn, depends upon the task the settlement pattern is to perform.

As an example, consider the following. Let us assume that three policies involving budgetary allocations are possible:

(1) concentrate investment in the early stages of development;
(2) concentrate investment in the later stages of development;
(3) invest equally each year;

and, that the total amount of money available under each policy is the same. In accord with this, figure 3 presents a possible distribution of investment over time for each of the policies, and the spatial distribution of settlements which might result from each policy. The problem now is to evaluate each of the three policies, and this involves consideration of the circumstances under which each policy would be optimal. Thus, policy 1 might be preferable if the planner can take only a short-run view, if location of rural settlements near Tel Aviv and other major cities is desirable (for example, if the provision of food to major urban centers is the chief function of rural places), or if money is available now which might not be available at a later date. Policy 2 might be preferable if protection against Arab belligerence or national expansion are important for developing certain areas. Policy 3 might be preferable if congestion is to be avoided or if the planner wants to create a setting in which future social change would be likely (made easier by mixing old settlements with new, whereas other budget policies result in a clustering of settlements of a single age).

Israeli policy implementation indicates that policies 2 and 3 have been developed by Israeli planning and settlement authorities. The implementation of policy 2 is apparent by the number of settlements which have emerged in militarily strategic areas, such as the Negev and the Dead Sea region. Often, these settlements are founded by pioneering youth groups or by paramilitary (*Nachal*-Pioneer-Defense Youth) groups of the Israeli Defense Forces who volunteer for rural settlement assignments. Examples of such settlements would be *Nachal Oz* along the Gaza Strip area, or *Yotvata* in the

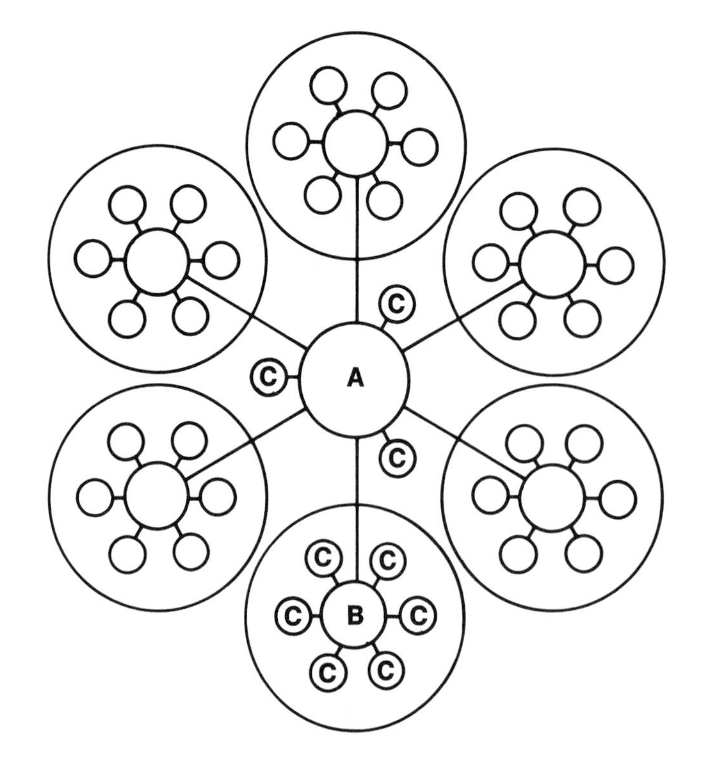


**A** Urban Center

**B** Village Group Center

**C** Village Settlement

Figure 4. Regional Plan Diagram, Lachish District.
Adapted from H. Darin-Drabkin. *Patterns of Cooperative Agriculture in Israel* (Tel Aviv: Israel Institute for Books, 1962), p. 216a.

southern Negev. Since the mid-1950s, regional development has taken on the form of policy 3; older settlements have been integrated into a regional system with newly established settlements, including central places (the "rural center") and the "regional town." These planning schemes were based on the creation of new settlements, mainly of the *moshav* type, which were often referred to as *moshav olim* (newcomer's *moshav* or immigrant *moshav).* These newer settlements were integrated with the various types of existing settlements to form the hinterland for the new central places. The central places were varied in size and function to form a hierarchy within the regional system (figure 4). An important function of this scheme was to facilitate the absorption and settlement of immigrants. As a general rule, immigrants having similar points of origin or cultural background were put in the same settlement.[13] Interaction with Israelis and immigrants from other countries would take place in the central place through its economic, educational, and social functions. This proved to be a rather successful approach for integrating the newcomers into the Israeli national culture, while at the same time maintaining the psychological security the immigrant had by living in the "old-country" atmosphere. Another important function of this planning scheme was to settle newcomers in the underpopulated areas of the country, rather than concentrating additional persons in the rapidly growing urban centers of Tel-Aviv, Haifa, and

13. H. Darin-Drabkin, op. cit., p. 204. See also, J. Shuval, *Immigrants on the Threshold* (Chicago: Atherton Press, 1963); and E. Cohen, L. Shamgar, and Y. Levy, *Absorption of Immigrants in a Development Town* (Jerusalem: Department of Sociology, Hebrew University, 1962).

Jerusalem. The Negev's *Lachish* area regional development plan, and the regional development schemes of the *Ta'anach* area in lower Galilee, the *Adullam* area in the Jerusalem Corridor, the *Korazim* area in the upper Jordan Valley, and the *Besor* area in the western Negev are implicit examples of this policy.

## CONCLUDING REMARKS

Contemporary planning and development of Israeli settlements is based on the assumption that rural planning should not be carried out at the level of individual villages. Rather, these plans should encompass a whole region consisting of a considerable number of villages, as well as semiurban central places. This policy and philosophy includes a role for a model and approach such as that presented in this paper. In the context of planning and development, these can serve as valuable tools for examining the effects of human and governmental behavior upon spatial patterns of Israeli rural settlement, as well as the interrelationships between a changing environment and a changing settlement pattern. Also, if the researcher has a well-defined idea of the task the system of settlements is to perform, the approach outlined here may also be used to provide guidelines for planning decisions, either through examining current situations or the effects of past decisions.

# 5

# ACTION SPACE DIFFERENTIALS IN CITIES

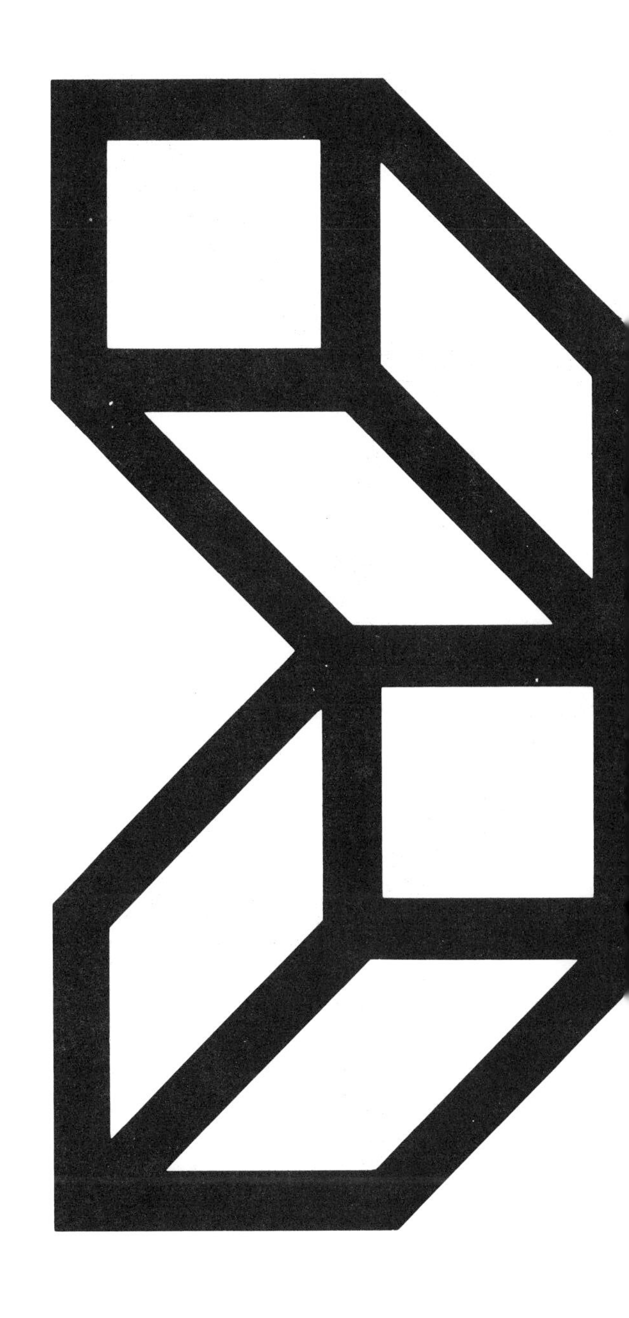

**Frank E. Horton**

**David R. Reynolds**
University of Iowa

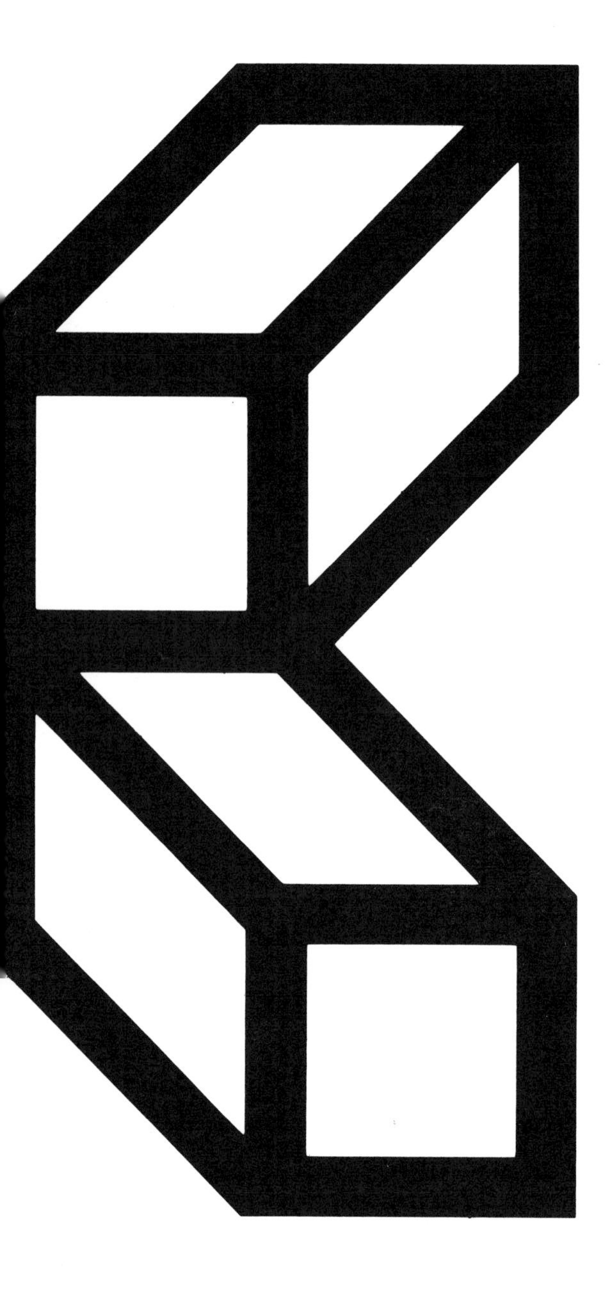

# Action-Space Differentials in Cities

**ABSTRACT**

The objective of the research set forth in this report is to identify the sources of variations in the action spaces of urban residents.* A conceptual basis for a model of action-space formation in the city is presented. Three empirical analyses are discussed—the spatial extent of action spaces in the city, the spatial structure of information surfaces, and the perception of residential quality—because they are necessary to articulate additional elements of a more complete model of action-space formation; so too are the implications of these analyses in relation to the aforementioned model.

* The support of the National Science Foundation and the Institute of Urban and Regional Research, University of Iowa, is gratefully acknowledged.

## IMPLICATION OF VARIATIONS IN ACTION SPACES OF URBAN RESIDENTS

The authors have previously defined *action space* as "the collection of all urban locations about which the individual has information and the subjective utility or preference he associates with these locations."[1] The action-space formation process can provide a basis for understanding several important urban spatial processes. The spatial structuring of urban areas is the result of a continuous reevaluation, in a very broad sense, by consumers and entrepreneurs. That is, the producers and distributors of goods and services in urban areas arrive at locational decisions based on imperfect information. The success or failure of the entrepreneur's operation is dependent on his long-range projection of consumer behavior decisions.

Since the individual as a consumer (*consumer* is here used in the broad sense of a person acting both as an employee and as a buyer or user of goods and services) cannot by his actions change the environment in which he is located, nor can he have an appreciable impact on the spatial structure of opportunities, he must decide where to acquire necessary goods and services on the basis of the opportunities available. Over the long run, therefore, some entrepreneurs are rewarded by continuous purchases or utilization of their goods or services, while others fail and drop out of the system. In the short run, the consumer continuously decides where he will acquire goods and services until such time as specific decisions become relatively habitual and entrenched.

The author's research into the action-space formation process, presupposing entrepreneurial decision-making, seeks only to identify the action spaces of urban residents through behavioral analysis of individuals at a specific location within an objective urban spatial structure. *Objective spatial structure* may be defined as "the actual locations of all potential activities and their associated objective levels of attractiveness within an urban area."[2] Thus, the individual is seen as responding to his knowledge and perception of an objective spatial structure. The individual's response to this structure is manifested in his *activity space,* "the subset of all urban locations with which the individual has direct contact as a result of day-to-day activities."[3]

The geographer, searching for order in spatial behavior and the spatial structure resulting from such spatial behavior, must examine, from both theoretical and empirical perspectives, the basic generative processes underlying spatial behavior. Action-space formation is one such process. As pointed out earlier, the individual acting as a member of a household is the basic behavioral entity that, in conjunction with all other urban

1. F. E. Horton and D. R. Reynolds, "Effects of Urban Spatial Structure on Individual Behavior," in *Papers Toward the Understanding of Change in Urban Spatial Structure,* ed. L. Brown and E. Moore, Economic Geography, Vol. 47, no. 1 (January 1971), pp. 36–48.

2. Ibid.

3. Ibid.

residents, plays an important role in the long-range spatial restructuring of our cities. Variation in the form and structure of our cities is the result of an interactive process between individuals and between individuals and their environments.

## A STAGE MODEL OF THE URBAN ENVIRONMENTAL LEARNING PROCESS

An individual's action space is the manifestation of his activity demands predicated upon his perception and cognition of urban locations which can potentially satisfy his demands. Neither an individual's perception of the urban environment nor his action space can be viewed as unchanging. Each is perhaps best viewed as an abstract component of a general environmental learning process. The nature of this process has not been investigated in any depth, although a few provocative articles pertaining to it have been published.[4] Based upon the work of Carr, the empirical findings of social psychology, and the preliminary findings of the author's on-going research into action-space formation, a tentative stage model of the learning process, at least for the newcomer to the city, can be presented.[5]

A generalized model of the hypothesized components and their linkages is presented in diagrammatic form in figure 1. These components and their relative impacts upon the learning process are discussed below. It is to be stressed that in this model it is assumed that the newcomer has already selected a residential site, even though in reality this decision itself is likely to be biased by his initial perception of his prospective urban environment. The several stages, as presented, are not viewed as discrete or mutually exclusive, but for any newcomer stage 1 initially is dominant over stage 2, and stage 3, while, for many individuals, increasing lengths of residence render stage 2 ever more dominant over stage 1. In the model, it is also assumed that the objective spatial structures of the urban environment do not change so rapidly as to frustrate attempts at learning.[6]

### Stage 1—Distance Bias

The initial spatial behavior patterns established by the individual are highly unstable except for the journey to work, which is likely to become routine quite soon. Other forms of intraurban spatial behavior are greatly influenced by (1) the individual's visual experience incurred on his way to work and that derived from interaction in his immediate residential area; (2) his set and level of activity demands; and (3) his travel preferences and his expectations of satisfaction at certain types of destinations, such as chain department stores and national food and gasoline chains, or more generally, the

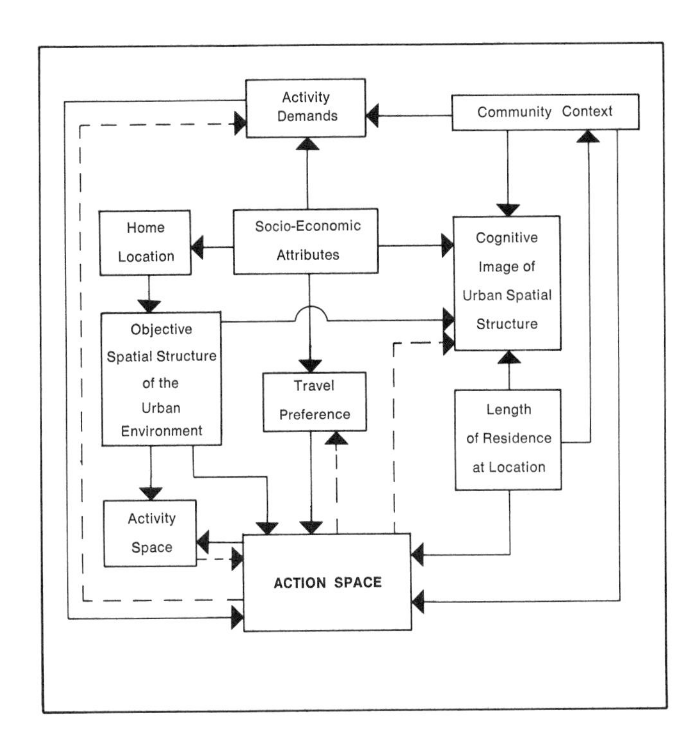


Figure 1. Conceptual Model of Action Space.

4. See, for instance, Stephen Carr, "City of the Mind," in, *Environment for Man,* ed, W. R. Ewald, Jr. (Bloomington: Indiana University Press, 1967) pp. 197–231.

5. This section of the paper represents an elaboration and extension of our conceptualization as presented in Horton and Reynolds, op. cit.

6. For a discussion of the likely impacts of rapid urban change on the process see Horton and Reynolds, op. cit.

central business district and planned shopping centers—both of these have been developed elsewhere, either in rural areas or in another urban place.

At this stage in the learning process, the individual's action space is expanding areally and his activity space at any moment in time is the result of his overt searching and is accordingly changing with the satisfactory fulfillment of his successive demands. The dominant foci in each of these spaces are the residence and work place, although at the same time other nodes are being sought and sampled by the individual and subjected to what may be termed spatial discounting procedure. Nodes proximate to the residence and work place are sampled first and stored as a component point in the individual's action space. As the individual satisfies demand at a more distant node, he subjectively compares it to those sampled and learned earlier. If the more distant node is perceived as being more satisfying to his demands, then the previously sampled nodes located nearer to either his residence or work place are discounted; that is, they are deleted from his current activity space but retained as component points in his action space, even though he no longer responds to them. It is through such behavior that the individual builds up his cognitive image of the urban environment and increases the size of his action space. The individual's spatial behavior is suboptimal, even though he may overtly be attempting to establish an action space in which he can fulfill his desires most satisfactorily with the least amount of additional travel.

### Stage 2—Community Socialization

By the beginning of the second stage, the individual has established a temporary equilibrium between his activities and the opportunities that he perceives within his action space. Up to this time, his direct and vicarious environmental experience that has been contributing to the development of his action space has been affected by distance and directional biases. He has now become, at least partially, integrated into informal social groups based upon either the residents of his immediate neighborhood or his associates at work. From these informal social contacts he learns of other potentially satisfactory destinations within the city. If these destinations are located within or near the perimeter of his action space, he is more likely to investigate their desirability—and again his previously satisfactory destinations are subjected to spatial discounting.

During this stage of the learning process, the relative importance of sources of environmental information shifts from direct contact to word-of-mouth communication with members of the social groups to which he belongs. Since a number of studies have demonstrated that there is close relationship between social and spatial propinquity,[7] it would appear that informal social contacts predicated upon residential location are the most important in providing environmental information for the new resident. The individual's spatial ordering of the urban environment now becomes more and more similar to that of other members of the community.[8] It is in this stage that the social, economic, and life-style characteristics of the individual and those dominant in his community are likely to exert a pronounced effect upon the learning process. Those individuals who are integrated most easily into a community or neighborhood and hence have increased sources of information are those whose social, economic, and life-style characteristics closely conform to the established community norms. For other individuals (for example, a factory foreman in a neighborhood in which professional people are predominant), membership in informal social groups may be denied with the result that they will probably lack this source of environmental information. Such an individual has, consequently, a higher probability of developing an environmental image and an action space which is at variance with those shared by comembers of informal community groups.[9] Although membership in both formal and informal community social groups fosters the development of shared environmental images, it may also inhibit the flow of information to the individual from other interpersonal sources. When such community cohesion occurs, the shared images are widely disparate from one community to another within the urban area. This, of course, would result in the neglect of some elements of the resident's objective spatial structure.

It is also in this stage that newspapers and other forms of mass media begin to exert more of an impact upon spatial behavior. In the distance bias stage too little of the urban environment was within the individual's action space and he rejected a major portion of the informational content of the mass media because of his insufficient locational information. Now, after partial integration into the community, his neighbors can help provide the relevant information that such media usually lack. The action space of an individual who is integrated into a community not only resembles those of other community members; it is also likely to be considerably more extensive and coherent than those of solitary residents.

### Stage 3—Spatial Equilibrium

At this stage, the individual's activity space is in equilibrium with his perception of opportunities. For many individuals the overt learning process is terminated, except for an occasional new piece of information, such as that coming from a change in the objective spatial structure (for example, the construction of a new

7. L. Festinger, S. Schacter, and K. Back, *Social Pressures in Informal Groups* (New York: Harper and Brothers, 1950); R. K. Merton, "The Social Psychology of Housing," in *Current Trends in Social Psychology* (Pittsburgh: University of Pittsburgh Press, 1950).

8. For evidence suggesting the existence of community-held images see, Kevin Lynch, *Image of the City,* (Cambridge, Mass. The M.I.T. Press, 1960).

9. In this regard, the work of some social scientists, sociologists, and social psychologists in particular, contain useful insights, albeit indirect, into the formation of an individual's action space and the extent to which action spaces are shared by similar groups of people. See, for example, A. H. Rubenstein and C. J. Haberstrol, *Some Theories of Organization* (Homewood, Ill.: Richard D. Irwin, 1964).

shopping plaza). For these individuals, spatial behavior has become routine and habitual; that is, their activity space remains stable. Therefore, any increase in the size of the action space is independent of the activity space; in fact, the action space is likely to become smaller as a result of personal disuse of formerly sampled and discounted opportunities within it. Since environmental learning takes place at a cost (expenditure of time, effort, and money), the income level of an individual is likely to be a determinant of the time duration of the process. For this reason, it can be hypothesized that low-income, central-city residents achieve such spatial equilibrium more quickly than other urban-area residents.

For others, the environmental learning process, although proceeding at a reduced rate unless there has been a change in the set and level of the demand for activities, continues largely as a result of their continued attention to information channels, both interpersonal and mass media.[10] The behavior of these individuals, provided that they do not experience a change in residence and that they do not perceive change in the environment itself, can be expected to approach so-called rational economic behavior as their action spaces become congruent with the total metropolitan area. The *extent* of the activity space is, by definition, less than the extent of the action space. The surface character of the activity space is due to the fact that many points and areas in the action space have for some reason been discounted. Further, the individual's initial expectations and preferences for travel may have changed as a result of his new environmental experience.

The empirical work recounted in the succeeding sections of this paper is focused upon several aspects of the above model of the learning process and its relationship to the formation of an action space. Specifically, our concern is with (1) community context and income differentials related to the extent of an individual's action space; (2) the relative prominence of apparent distance biases in forming the composite action spaces of urban residents in two disparate community contexts; and (3) the effect of income and location on an individual's idea of residential quality. Owing to problems of data access our approach is of necessity cross-sectional rather than longitudinal.

## AN EMPIRICAL ASSESSMENT OF VARIATION IN ACTION SPACES

### The Data Base

The data utilized in the research were acquired through home interviews in Cedar Rapids, Iowa, a city with a population of approximately 120,000. Information was gathered from individuals in urban residential areas

10. Changes in the demand for activities may be induced by changes in the life cycle of the individual or by changes in the supply of activities present in the environment.

widely disparate in terms of job locations and shopping opportunities, predominant life styles and residential preferences. Two areas in Cedar Rapids were selected: one, a recently developed, middle class, suburban community located on the urban fringe and having a very homogeneous residential quality; and the second, an old, low-income, central-city area with highly variable, yet generally low, residential quality. For brevity, the suburban area will be referred to as Cedar Hills and the central-city area as Oak Hill. A profile of the two samples is presented in table 1.

Table 1. *Socioeconomic Profile* of the Two Samples

| | Oak Hill-Jackson | Cedar Hills |
|---|---|---|
| Dominant occupational category | Mfg.-Oper. | Prof.-Tc. |
| Median Family Size | 3 | 4 |
| Median School Years | 11 | 1 year college |
| Median Family Income | $6–8,000 | $10–14,000 |
| Median Length of Residence | 4 | 3 |

### Operational Definitions

Action space, the basic concept of interest in this investigation, is characterized by two components: first, a set of locations which define its spatial extent; and second, a varying surface specifying the level of information possessed by the individual for each location. To obtain an operationally useful definition of action space and "perception of residential quality," each respondent was confronted with a large-scale map of the Cedar Rapids metropolitan area on which twenty-seven residential areas were delimited. The respondent was asked to indicate on a five-point scale his level of familiarity with each of the areas delimited. For each area with which the respondent was familiar to some degree, he was also asked to express his perception of its residential quality; again on a five-point scale this time of goodness, from very poor to very good. Since several of our analyses demanded interval measurement, the ordinal-familiarity responses and housing-quality responses were transformed to an interval scale by the application of a multidimensional scaling technique derived from Thurstone's Law of Categorical Judgment.[11]

Given these data, the extent of an individual's action space was simply defined as the number of residential areas with which he had some degree of familiarity. The *form* of the action space was operationally defined as the zero order correlations between residential areas of varying scaled levels of perceived familiarity. With this type of information available, it was possible to determine the most salient or basic structural dimensions of the aggregate action spaces of any subgroup of respondents by employing factor analytic procedures.

### Spatial Extent of Action Spaces

With regard to the spatial extent of action spaces we wished to test two hypotheses:

(1) That the extent of the aggregate action space of the Cedar Hills sample was greater than that of the Oak Hill sample; and

(2) That within each of the two samples the higher the income level of an individual, the greater the extent of the action space.

11. See Horton and Reynolds, op. cit.

Table 2. *Chi-Square Analysis* of Unfamiliarity

| | Frequency of Unfamiliarity | | | |
|---|---|---|---|---|
| | Cedar Hills | | Oak Hill | |
| Residential Region | Observed | Expected | Observed | Expected |
| 1 | 370 | 355.8 | 430 | 444.2 |
| 2 | 148 | 143.6 | 175 | 179.4 |
| 3 | 289 | 236.6 | 243 | 295.4 |
| 4 | 257 | 257.0 | 321 | 321.0 |
| 5 | 240 | 205.9 | 223 | 257.1 |
| 6 | 134 | 99.2 | 89 | 123.8 |
| 7 | 154 | 293.9 | 507 | 367.1 |

$\chi^2 = 174.316$

$\chi^2_6(.01) = 16.812$

Table 3. *Chi-Square of Unfamiliarity* by Income and Residential Region—Cedar Hills Sample

| | Frequency of Unfamiliarity Income Category | | | | | | | | | |
|---|---|---|---|---|---|---|---|---|---|---|
| | L.T. $6,000 | | $6–7,999 | | $8–9,999 | | $10–13,999 | | G.T. $14,000 | |
| Residential Region | Obs. | Ex. | Obs. | Ex. | Obs. | Ex. | Obs. | Ex. | Obs. | Ex. |
| 1 | 21 | 16.3 | 52 | 47.7 | 102 | 116.0 | 162 | 162.0 | 33 | 27.7 |
| 2 | 9 | 6.5 | 17 | 19.1 | 47 | 46.4 | 61 | 64.8 | 14 | 11.1 |
| 3 | 10 | 12.7 | 36 | 37.2 | 80 | 90.6 | 133 | 126.5 | 21 | 21.6 |
| 4 | 9 | 11.3 | 35 | 33.1 | 82 | 80.6 | 113 | 112.5 | 16 | 19.1 |
| 5 | 9 | 10.6 | 28 | 30.9 | 77 | 75.2 | 107 | 105.1 | 19 | 17.9 |
| 6 | 6 | 5.9 | 15 | 17.3 | 47 | 42.0 | 58 | 58.7 | 8 | 10.0 |
| 7 | 6 | 6.8 | 22 | 19.8 | 55 | 48.3 | 63 | 67.4 | 8 | 11.5 |

$\chi^2 = 13.287$

$\chi^2_2\,(.05) = 13.848$

Table 4. *Chi-Square of Unfamiliarity* by Income and Residential Region—Oak Hill Sample

| | Frequency of Unfamiliarity Income Category | | | | | | | | | |
|---|---|---|---|---|---|---|---|---|---|---|
| | L.T. $3,000 | | $3–5,999 | | $6–7,999 | | $8–9,999 | | $10–13,999 | |
| Residential Region | Obs. | Ex. | Obs. | Ex. | Obs. | Ex. | Obs. | Ex. | Obs. | Ex. |
| 1 | 80 | 77.6 | 124 | 120.7 | 118 | 121.1 | 78 | 81.1 | 30 | 29.4 |
| 2 | 27 | 31.6 | 44 | 49.1 | 52 | 49.3 | 37 | 33.0 | 15 | 12.0 |
| 3 | 37 | 43.9 | 62 | 68.2 | 78 | 68.5 | 49 | 45.8 | 17 | 16.6 |
| 4 | 61 | 58.0 | 90 | 90.5 | 95 | 90.4 | 54 | 60.6 | 21 | 22.0 |
| 5 | 39 | 40.3 | 66 | 62.6 | 63 | 62.8 | 39 | 42.1 | 16 | 15.3 |
| 6 | 16 | 16.1 | 24 | 25.0 | 26 | 25.1 | 19 | 16.8 | 4 | 6.1 |
| 7 | 99 | 91.6 | 148 | 142.3 | 128 | 142.8 | 99 | 95.6 | 33 | 34.7 |

$\chi^2 = 11.430$

$\chi^2_{24}\,(.02) = 11.992$

It was anticipated that the aggregate action space of the Cedar Hills samples would be greater, partly because of the generally higher socioeconomic status of this group and its peripheral location in the metropolitan area which necessitates considerable travel to go to work and to shop. The second hypothesis was derived from the simple notion that there is a direct relationship between increasing income and the need for increasingly diverse spatial behavior.

In order to test the first hypothesis, the twenty-seven residential areas were combined into seven larger residential regions (figure 2) based on the similarity of such objective residential characteristics as income and median housing value. This was necessary so that the problem of very low frequencies could be avoided. Then, the frequency with which Oak Hill and Cedar Hills respondents indicated they were unfamiliar with the residential areas comprising each of these regions was recorded forming the 7 × 2 contingency table displayed as table 2. The null hypothesis of no difference between the two samples in terms of the frequency of unfamiliarity with residential areas was assessed by the chi-square test and rejected at the 0.01 level of significance. The cell entries in table 2 are such that it may be concluded that the extent of the aggregate action space of the Cedar Hills sample was significantly greater than that of the Oak Hill sample.

The within-sample income variation hypothesis was tested by first grouping each respondent into one of five mutually exclusive income categories. (The categories of the two samples differed slightly because of the differing ranges of income in the two areas.) Tables 3 and 4 are the two within-sample contingency tables formed by income category which indicate the frequency of *unfamiliarity* with the residential areas comprising each of the seven regions described earlier. The null hypothesis that the frequency of unfamiliarity with residential areas is independent of income was examined by the chi-square test and was not rejected even at very low levels of significance. The research hypothesis is, therefore, rejected and it must be concluded that, within the residential areas under examination, variation in the extent of action spaces is not attributable to variations in income. This finding may be due in part to the relatively small size of the Cedar Rapids metropolitan area, but the fact that the research hypothesis was accepted suggests that residential location in the metropolitan area is a more pervasive determinant of the extent of an action space.

## Spatial Structure of Action Spaces

There are a number of information channels open to the individual through which he acquires urban environ-

Figure 2. The Residential Areas of Cedar Rapids Evaluated by Respondents as Grouped into "Regions."

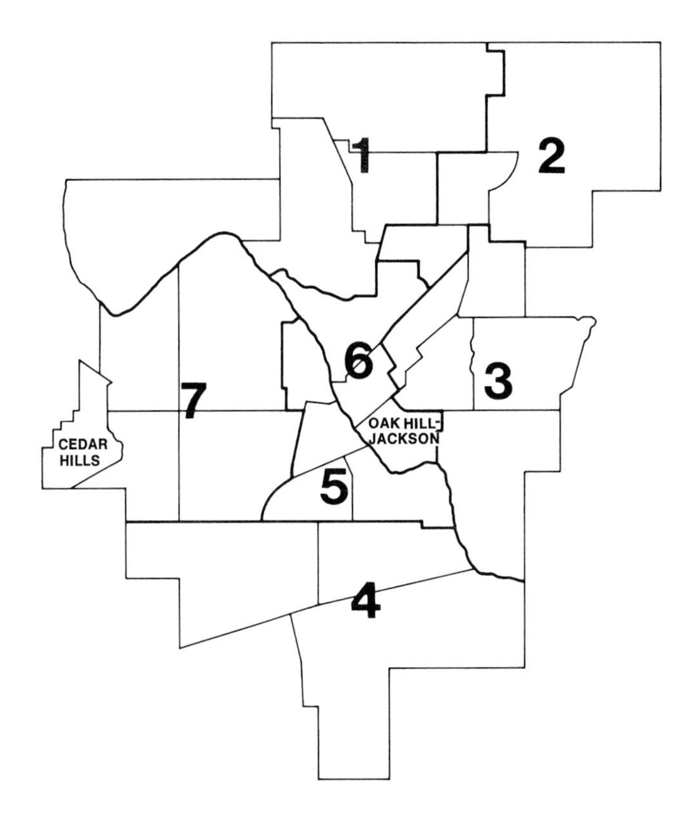


mental information and becomes aware of the residential areas in the city:

(1) Direct contact via past and present day-to-day activities;
(2) Friends and social acquaintances; and
(3) The mass media, etc.

The little empirical information there is available, however, suggests that direct personal contact is the most salient mode of urban environmental learning. Since the data indicated that there was a high degree of intraurban residential mobility in both the Oak Hill and Cedar Hills samples, it was apparent that the basic structures of aggregate action spaces will be partially obscured by information acquired at previous residences in the metropolitan area.

In an attempt to establish a statistical control for an individual's level of familiarity with residential areas attributable to his previous residential location, each of the two samples were subdivided into groups of at least ten individuals who had resided in the same residential area prior to moving to their present residence. Thus, for the Cedar Hills sample, ten groups were formed, one of which was comprised of individuals who had no direct residential experience in an area other than Cedar Hills. Eight groups were formed from the Oak Hill sample.

The spatial structure of the aggregate action space for each of the two samples was examined in the follow-

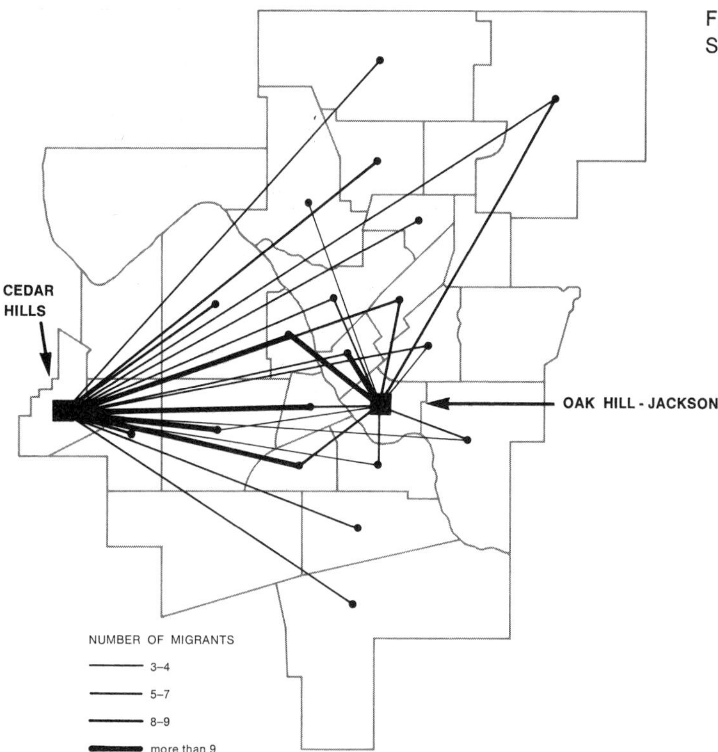


Figure 3. Location of Previous Areas of Residence for the Two Samples.

Table 5. *Aggregate Structure* of the Cedar Hills Action Space Resulting from the Varimax Factor Structure (for those with previous residential experience in Cedar Rapids)

| | Loadings on Each Factor | | | | |
|---|---|---|---|---|---|
| Sub-Area | 1 | 2 | 3 | 4 | 5 |
| 1 | | | | | −0.612 |
| 2 | | | | | −0.832 |
| 3 | | | −0.746 | | |
| 4 | | | −0.533 | | |
| 5 | | −0.531 | | | |
| 6 | | −0.745 | | | |
| 7 | | −0.780 | | | |
| 8 | | −0.409 | | | |
| 9 | 0.607 | | | | |
| 10 | | | | 0.438 | |
| 11 | | | | | −0.461 |
| 12 | | −0.463 | | | |
| 13 | | −0.644 | | | |
| 14 | | −0.619 | | | |
| 15 | | −0.645 | | | |
| 16 | | | | 0.750 | |
| 17 | | | | 0.742 | |
| 18 | | | | 0.566 | |
| 19 | 0.720 | | | | |
| 20 | | −0.549 | | | |
| 21 | 0.650 | | | | |
| 22 | 0.546 | | | | |
| 23 | 0.763 | | | | |
| 24 | 0.792 | | | | |
| 25 | 0.741 | | | | |
| 26 | 0.631 | | | | |
| 27 | 0.661 | | | | |
| % of Variance | 41.4 | 7.6 | 5.0 | 4.8 | 4.0 |

ing two-step manner. First, a principle components factor analysis of the between-area correlation matrix of scaled familiarity *uncorrected* for previous residence was conducted; then a principle components factor analysis of the "within group," between-area correlation matrix of scaled familiarity. This "within group" correlation matrix was developed by performing covariance adjustments on each of the previous areas of residence groups. In this way, it was possible to remove statistically from the action space those unique levels of familiarity attributable to the location of previous residence.

For the Cedar Hills sample it was hypothesized that the two resulting factor structures (describing the structure of the action space) would differ to a marked extent as a result of the wide dispersion of previous areas of residence for many of the respondents (see figure 3). With regard to the Oak Hill sample, it was assumed that the two resulting factor structures would be almost identical since the previous areas of residence were spatially concentrated near the central portion of the city.

The results of the analysis tended not only to confirm our hypotheses; they proved to be interesting in their own right. From the principle components factor analysis of the Cedar Hills sample between-area familiarity correlations uncorrected for area of previous residence, five basic dimensions of the aggregate action space were extracted (see table 5). In this and subsequent factor analyses, the resulting factor structures were characterized by each of the twenty-seven areas having a high factor loading on one and usually only one of the basic

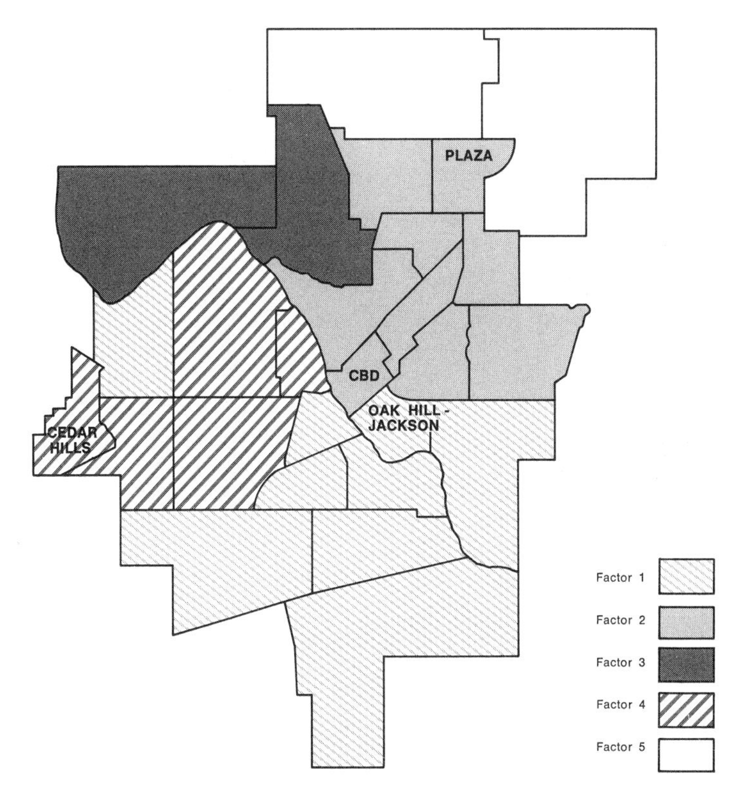


Figure 4. "Uncorrected" Familiarity Factor Structure—Cedar Hills Sample.

dimensions. For this reason, it is possible to present our results in map form. The five dimensions depicted on figure 4 account for 63 per cent of the between-area levels of familiarity. The major dimension of variation in between-area familiarity is accounted for by variations within area 1, which can generally be characterized by decreasing familiarity as it stretches in a southerly direction from the central city. The remaining dimensions of variation in decreasing order of "explained" variation may be characterized as follows: (1) an area along the major transportation axis of the city characterized by a surface of familiarity peaking at the CBD (central business district) and major shopping plaza and approximately uniform out from this axis, (2) a relatively inaccessible, low-population-density area of very low levels of familiarity, (3) an area of decreasing familiarity out from the home area and towards the CBD, and (4) a discontinuous area of mixed levels of familiarity.

The principle components factor analysis on the Cedar Hills correlation matrix corrected for area of previous residence resulted in 6 dimensions of the aggregate action space (see table 6). These account for 66 percent of the corrected, between-area levels of familiarity. These are presented in figure 5. The major dimension of variation is areally similar to that of the preceding analysis, except that all the component residential areas comprising it are now contiguous. In fact, there are no spatial discontiguities in any of the basic dimensions of between-area familiarity resulting from this analysis. Factor 2 is an area of low levels of familiarity; factor 3 is similar

Table 6. *Aggregate Structure* of the Cedar Hills Action Space Corrected for Previous Residence Resulting from the Varimax Factor Structure

| | Loadings on Each Factor | | | | | |
|---|---|---|---|---|---|---|
| Sub-Area | 1 | 2 | 3 | 4 | 5 | 6 |
| 1 | | 0.639 | | | | |
| 2 | | 0.821 | | | | |
| 3 | | | | | 0.815 | |
| 4 | | | | | 0.563 | |
| 5 | | | −0.456 | | | |
| 6 | | | −0.781 | | | |
| 7 | | | −0.791 | | | |
| 8 | | | −0.500 | | | |
| 9 | −0.597 | | | | | |
| 10 | | | | | | −0.602 |
| 11 | | | | | | −0.501 |
| 12 | | | −0.515 | | | |
| 13 | | | −0.630 | | | |
| 14 | | | −0.524 | | | |
| 15 | | | −0.634 | | | |
| 16 | | | | 0.775 | | |
| 17 | | | | 0.730 | | |
| 18 | | | | 0.529 | | |
| 19 | −0.722 | | | | | |
| 20 | | | | | | 0.536 |
| 21 | −0.629 | | | | | |
| 22 | −0.502 | | | | | |
| 23 | −0.792 | | | | | |
| 24 | −0.821 | | | | | |
| 25 | −0.722 | | | | | |
| 26 | −0.646 | | | | | |
| 27 | −0.662 | | | | | |
| % of Variance | 40.8 | 7.2 | 5.2 | 4.8 | 4.0 | 3.8 |

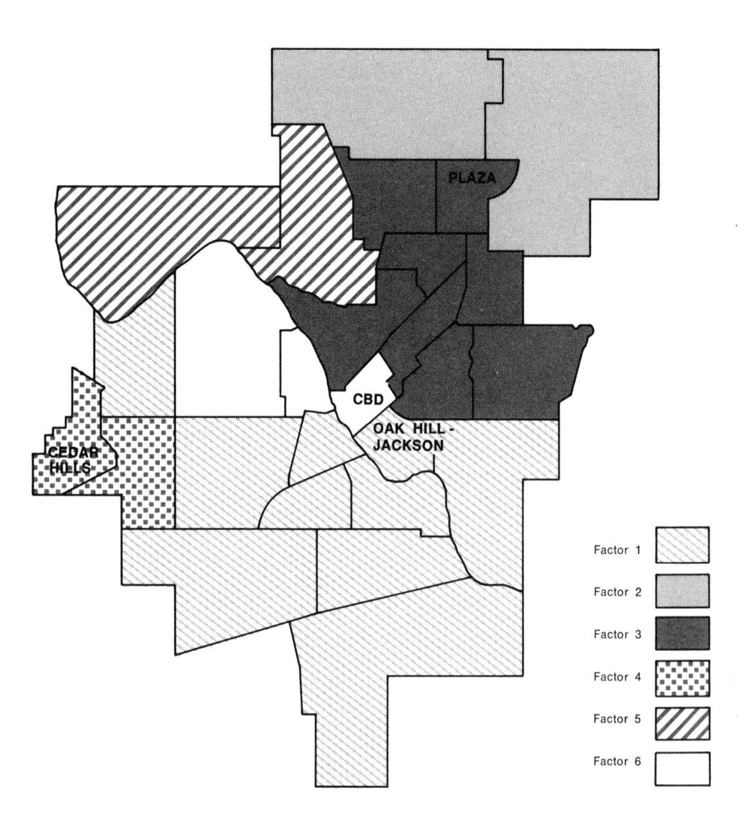


Figure 5. "Corrected" Familiarity Factor Structure—Cedar Hills Sample.

Table 7. *Aggregate Structure* of the Oak Hill Action Space Corrected for Previous Residence Resulting from the Varimax Factor Structure

| | Loadings on Each Factor | | | | |
|---|---|---|---|---|---|
| Sub-Area | 1 | 2 | 3 | 4 | 5 |
| 1 | −0.457 | | | | |
| 2 | −0.497 | | | | |
| 3 | −0.741 | | | | |
| 4 | −0.595 | | | | |
| 5 | −0.701 | | | | |
| 6 | | | −0.560 | | |
| 7 | −0.659 | | | | |
| 8 | −0.572 | | | | |
| 9 | | | | −0.729 | |
| 10 | | | | | −0.708 |
| 11 | | | | | −0.797 |
| 12 | | | | −0.548 | |
| 13 | | | | −0.564 | |
| 14 | | | −0.784 | | |
| 15 | | | −0.723 | | |
| 16 | | | | −0.723 | |
| 17 | | | | −0.815 | |
| 18 | | | | | −0.668 |
| 19 | | | | | −0.550 |
| 20 | | | −0.662 | | |
| 21 | | 0.681 | | | |
| 22 | | 0.736 | | | |
| 23 | | | | | −0.512 |
| 24 | | 0.599 | | | |
| 25 | | | | −0.553 | |
| 26 | | 0.495 | | | |
| 27 | | | | −0.566 | |
| % of Variance | 40.4 | 7.1 | 6.1 | 5.3 | 4.4 |

to factor 2 of the preceding analysis except that it no longer includes the CBD; factor 4—a neighborhood factor—is similar to the previous factor 4 except that it is reduced in areal size; factor 5 is identical to the previous factor 3; and factor 6 is a new dimension of decreasing familiarity to the west of the CBD.

In general, the results of these two analyses indicate that the action space varies systematically over the urban area. A decline in familiarity corresponding to the distance from the principal interaction areas—the neighborhood, the CBD, and the major shopping plaza—appears to be the dominant spatial regularity. Furthermore, these regularities appear more pronounced when the locations of previous areas of residence have been statistically controlled.

The same two-step procedure was employed in the principle components factor analyses conducted for the Oak Hill sample. The analysis based on within-group, between-area familiarity resulted in a factor structure (see table 7) indistinguishable from that based on "uncorrected" between-area familiarity (see table 8), even in terms of the total explained variance (63 per cent in both). This is in conformity with our hypothesis.

As can be seen from figure 6, the aggregate action space of the Oak Hill sample differs markedly from that of the suburban sample. The spatial dimensions of variations in familiarity in decreasing order of "explained" variation are: (1) a large northern area of uniform unfamiliarity (with the single exception of the area containing the major shopping plaza; (2) a fairly large area of

Figure 6. Familiarity Factor Structure—Oak Hill Sample.

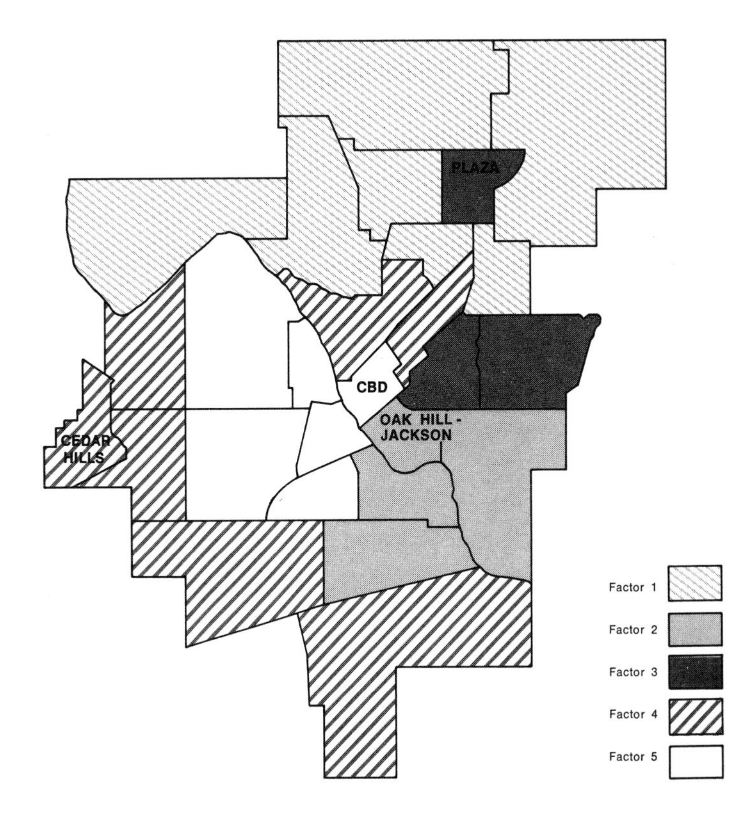


high familiarity levels focusing on the home area (it is interesting to note that the residential areas comprising this are the so-called old Bohemian residential areas); (3) a discontiguous area of moderate familiarity comprised of the highest-income residential areas in the city; (4) a more or less concentric ring of peripheral residential areas of low levels of familiarity but not as low as those in the northern part of the metropolitan area; and (5) an area of decreasing familiarity west of the CBD.

Again, it is evident that the aggregate action space of the central-city sample is much more spatially concentrated than that of the suburban sample. More importantly, however, familiarity with residential areas appears to be more a function of the social and economic characteristics of the areas than it is in the peripheral suburban area. The structure of action space for residents in the latter area appears to be dominated more by direct contact incurred in the pursuit of such activities as going to work and shopping.

### Perception of Residential Quality

Having determined that the extent and structure of the action spaces for the two sets of respondents differed significantly, the next step in our investigation was to ascertain the extent to which the overall perceptions of residential quality in Cedar Rapids differ between and within the two sets and the extent to which such differences (if apparent) are locationally biased; that is, concentrated in only a few residential areas. In this regard, five hypotheses were formulated and tested. These hy-

Table 8. *Aggregate Structure* of the Oak Hill Action Space Resulting from the Varimax Factor Structure (for those with previous residential experience in Cedar Rapids)

| | Loadings on Each Factor | | | | |
|---|---|---|---|---|---|
| Sub-Area | 1 | 2 | 3 | 4 | 5 |
| 1 | 0.451 | | | | |
| 2 | 0.481 | | | | |
| 3 | 0.724 | | | | |
| 4 | 0.587 | | | | |
| 5 | 0.720 | | | | |
| 6 | | | −0.541 | | |
| 7 | 0.652 | | | | |
| 8 | 0.589 | | | | |
| 9 | | | | −0.730 | |
| 10 | | | | | 0.658 |
| 11 | | | | | 0.740 |
| 12 | | | | −0.553 | |
| 13 | | | | −0.565 | |
| 14 | | | −0.784 | | |
| 15 | | | −0.725 | | |
| 16 | | | | −0.711 | |
| 17 | | | | −0.824 | |
| 18 | | | | | 0.640 |
| 19 | | | | | 0.508 |
| 20 | | | −0.685 | | |
| 21 | | 0.712 | | | |
| 22 | | 0.728 | | | |
| 23 | | | | | 0.466 |
| 24 | | 0.586 | | | |
| 25 | | | | −0.569 | |
| 26 | | 0.502 | | | |
| 27 | | | | −0.588 | |
| % of Variance | 40.8 | 7.1 | 6.0 | 4.9 | 4.2 |

potheses and their underlying rationale are discussed below.

**Hypothesis 1:**

Taken over all twenty-seven residential areas, the central-city respondents perceive the residential quality of the metropolitan area to be higher than those in the suburban area.

*Rationale.* Since the Oak Hill respondents are by definition living in the lowest-income area of the city and are likely to be cognizant of this fact, they are also those most likely to view other areas as better than they may actually be. Contrariwise, the Cedar Hills residents, since they reside in the residential area farthest from the CBD, are more likely to have considered and rejected alternative residential areas in arriving at their individual relocation decisions. Hence, they are likely to have a more accurate evaluation of alternative residential areas. (Recall that it has already been established that the respondents in Cedar Hills are familiar with more areas than are those in Oak Hill.)

**Hypothesis 2:**

The two sets of respondents will agree in their aggregate rankings of residential areas by quality only in terms of those at the extremes of a "goodness" continuum—that is, they will agree about which areas are the highest and lowest quality, but disagree on the rankings of those areas of intermediate residential quality.

*Rationale.* The rationale behind this hypothesis is the same as that for the preceding. It was anticipated that the Oak Hill respondents would know what areas are similar to their own in terms of residential quality and would also have received enough environmental information to have decided which are the best in the city, but that they would be indifferent between these two extremes.

**Hypotheses 3 and 4:**

There are no significant differences in the perceptions of residential quality in the metropolitan area between respondents with varying incomes in the suburban area, but such differences will be manifest for the central-city respondents.

*Rationale:* The rationale underlying the differences in these hypotheses is as follows. Cedar Hills is a recently developed subdivision in the city; hence, none of the homes are old and the variability in their assessed valuation is extremely low. It is safe to assume, therefore, that the residential quality of the area is homogeneous. Furthermore, it is unlikely that the householders have chosen to reside there for reasons other than similar personal preferences. Given the location of employment and shopping opportunities, it does not appear plausible that any residents have selected Cedar Hills as an area of residence for reasons of accessibility. As a result, constraints such as income should not be expected to vary with perceptions of housing quality.

Oak Hill, on the other hand, by virtue of its central location and heterogeneous housing conditions, is likely to contain residents who have chosen to live there for a plethora of reasons—the three principal ones are low rent, ethnicity, and accessibility. The lower-income families probably reside there from necessity, as a result of their low incomes; whereas the higher-income families are there due to choice. As a result, the lower-income respondents are more likely than their higher-income neighbors to consider other residential areas to be of a higher quality for the reasons stated previously.

**Hypothesis 5:**

The primary differences in the perception of residential quality between individuals with differing income levels in the central-city area will pertain to those areas in close proximity to the neighborhood.

*Rationale:* This hypothesis is derived from the psychological notion that for low-income residents who live in an area by necessity rather than choice, small differences in the objective characteristics of residential areas loom rather large in terms of their subjective perceptions of these characteristics. Since the areas proximate to Oak Hill are of only a slightly higher quality (in terms of income and median housing value), these are the areas in which variations by income in the perception of residential quality should be most pronounced.

**Hypothesis Testing**

The first hypothesis—that the Oak Hill respondents perceive the residential quality of Cedar Rapids as a whole to be higher than those in Cedar Hills—was tested by forming a contingency table (see Table 9) indicating, for each sample, the total frequency with which residential areas received an ordinal quality evaluation of 0, 1, 2, 3, and 4. This is the form in which the residential quality perceptions were obtained. Table 9 was tested for column independence by calculating the chi-square statistic, thereby testing the hypothesis of differences in the perceptions of the two samples. The difference between the two samples was found to be significant statistically at the 0.01 level; the cell entries verified that the Oak Hill respondents were indeed prone to evaluate the residential quality of the metropolitan area higher than were the Cedar Hills respondents.

The second hypothesis, concerning the aggregate rank ordering of the twenty-seven residential areas of the two samples, was tested in the following manner. The mean of the intervally scaled residential quality responses was calculated for each of the twenty-seven

Table 9. *Chi-Square Analysis* of Overall Residential Quality

| | Frequency of Residential Quality Response | | | |
|---|---|---|---|---|
| | Cedar Hills | | Oak Hill | |
| Residential Quality Response | Observed | Expected | Observed | Expected |
| 0 | 408 | 365.0 | 269 | 312.0 |
| 1 | 570 | 551.1 | 452 | 470.9 |
| 2 | 1235 | 1149.0 | 896 | 982.0 |
| 3 | 1166 | 1260.1 | 1171 | 1076.9 |
| 4 | 582 | 635.7 | 597 | 543.3 |

$\chi^2 = 51.456$
$\chi_3\,(.01) = 11.341$

Table 10. *Aggregate Evaluation* of the Quality of Residential Areas

| | Oak Hill Sample | | Cedar Hills Sample | |
|---|---|---|---|---|
| Area | Mean of Scaled Residential Quality | Rank | Mean of Scaled Residential Quality | Rank |
| 1 | 0.498 | 16 | 0.382 | 15 |
| 2 | 0.542 | 15 | 0.514 | 13 |
| 3 | 0.757 | 9 | 0.677 | 9 |
| 4 | 0.897 | 6 | 0.814 | 6 |
| 5 | 1.013 | 3 | 0.791 | 7 |
| 6 | 0.792 | 8 | 0.580 | 12 |
| 7 | 0.711 | 11 | 0.668 | 10 |
| 8 | 1.512 | 1 | 1.963 | 1 |
| 9 | 0.710 | 12 | −0.021 | 20 |
| 10 | 0.575 | 14 | 0.154 | 16 |
| 11 | −0.446 | 26 | −0.775 | 25 |
| 12 | 0.119 | 22 | −0.001 | 19 |
| 13 | 0.324 | 18 | 0.073 | 17 |
| 14 | 0.959 | 5 | 1.163 | 4 |
| 15 | 1.448 | 2 | 1.635 | 2 |
| 16 | 0.987 | 4 | 1.198 | 3 |
| 17 | 0.729 | 10 | 1.061 | 5 |
| 18 | 0.655 | 13 | 0.613 | 11 |
| 19 | −0.436 | 25 | −0.744 | 24 |
| 20 | −0.399 | 24 | −0.909 | 26 |
| 21 | −0.570 | 27 | −1.082 | 27 |
| 22 | 0.428 | 17 | 0.427 | 14 |
| 23 | 0.304 | 19 | −0.105 | 21 |
| 24 | −0.060 | 23 | −0.335 | 22 |
| 25 | 0.198 | 21 | 0.029 | 18 |
| 26 | 0.884 | 7 | 0.753 | 8 |
| 27 | 0.258 | 20 | −0.450 | 23 |

Rank Order Correlation = 0.943

areas for each sample. The areas were then ranked from highest to lowest in residential quality on the basis of the means for each sample (see table 10). The ranking was then compared by computing Spearman's rank order correlation coefficient. The correlation between the rankings of the two groups was an amazingly high 0.943. This forced the rejection of our research hypothesis, since there appeared to be general agreement between the two samples on a ranking continuum from best to poorest in terms of residential quality, even though the Cedar Hills respondents perceived a greater absolute disparity from the highest quality area to the lowest, and the Oak Hill respondents in general perceived individual areas to be of higher absolute quality. This finding is particularly interesting since it suggests that the two samples perceive residential quality on the same general psychological continuum; yet all areas are shifted up or down on the scale depending upon which sample one is considering.

The third and fourth hypotheses, it should be recalled, pertained to differences in the perception of residential quality based upon income differences. These were tested by developing a contingency table in which the cell entries indicated the frequency with which respondents in the five different income categories described earlier evaluated residential areas on the ordinal scale of 0, 1, 2, 3, and 4. Two such income category by evaluation category matrices were developed—one for each of the two samples (see tables 11 and 12). A chi-square test of the contingency table for the Cedar Hills sample indicated, in conformance with the third hypothesis, no significant difference between income category and the perception of residential quality even at very low levels of significance. A chi-square test of the contingency table for the Oak Hill sample, however, indicated that there were statistically significant differences at the 0.01 level in the perception of housing quality between respondents with different income levels. Furthermore, the cell frequencies are also in agreement with the research hypothesis. Respondents in the lowest-income category (income $3,000 or less) have the greatest tendency to perceive areas as high in quality and the lowest tendency to perceive areas as very low in residential quality.

The fifth and final hypothesis—that differences in the perception of residential quality between respondents in the Oak Hill sample with differing income levels will pertain to the perceptions of residential areas near the home area—was tested by calculating chi-square statistics for seven contingency tables. For each of the seven regions utilized in the analysis of the extent of action spaces, a separate table was formed. Each was of dimension 5 x 5 and indicated the frequencies with which the areas comprising the region in question were evaluated on the 0,

1, 2, 3, 4 residential quality scale by respondents in each of the five different income categories. Significant differences at the 0.05 and 0.02 levels were found in the perceptions of residential quality between respondents of varying income levels for the areas comprising region 1 (see figure 2) in the northern part of the city and for those comprising region 5 in the central city, respectively. Little credence can be placed in the differences found between respondents of various income levels for region 1 due to the low frequencies in the table. (Recall that the unfamiliarity of the Oak Hill respondents was concentrated in the northern part of the city.) However, the differences in perceptions of the residential quality of those areas proximate to Oak Hill—namely those found in region 5 (see table 13)—are particularly interesting. Contrary to our hypothesis, it is not the lowest-income group (under $3,000) which perceives the quality to be inordinately high in these areas but those individuals in the two lower intermediate-income categories who do so—namely those individuals with incomes of $3,000–6,000 and $6,000–8,000. At the present time we have no real explanation for this phenomenon, although we suspect that the relationship between income and perception is being confounded by other more salient social factors, particularly ethnicity. Therefore, with regard to this hypothesis, we must tentatively conclude that there are indeed spatial biases in the perception of residential quality, particularly those likely to be based on distance, but these do not vary as a simple function of income alone.

Table 11. *Chi-Square Analysis* of Perceived Residential Quality by Income Category—Cedar Hills Sample

| Income | Residential Quality Response | | | | | | | | | |
|---|---|---|---|---|---|---|---|---|---|---|
| | 0 | | 1 | | 2 | | 3 | | 4 | |
| Category | Obs. | Ex. | Obs. | Ex. | Obs. | Ex. | Obs. | Ex. | Obs. | Ex. |
| L.T. $6,000 | 17 | 15.0 | 20 | 21.0 | 44 | 45.5 | 42 | 43.0 | 23 | 21.5 |
| $6–7,999 | 36 | 50.1 | 66 | 71.2 | 147 | 151.5 | 160 | 143.1 | 77 | 71.4 |
| $8–9,999 | 139 | 132.2 | 186 | 184.6 | 400 | 400.0 | 383 | 377.7 | 175 | 188.5 |
| $10–13,999 | 189 | 175.7 | 252 | 245.5 | 525 | 531.9 | 486 | 502.2 | 254 | 250.6 |
| G.T. $14,000 | 27 | 35.0 | 46 | 48.9 | 119 | 106.0 | 95 | 100.1 | 53 | 50.0 |

$\chi^2 = 14.685$

$\chi^2_{16}(.01) = 32.00$

Table 12. *Chi-Square Analysis* of Perceived Residential Quality by Income Category—Oak Hill Sample

| Income | Residential Quality Response | | | | | | | | | |
|---|---|---|---|---|---|---|---|---|---|---|
| | 0 | | 1 | | 2 | | 3 | | 4 | |
| Category | Obs. | Ex. | Obs. | Ex. | Obs. | Ex. | Obs. | Ex. | Obs. | Ex. |
| L.T. $3,000 | 58 | 55.2 | 97 | 92.7 | 202 | 183.7 | 216 | 240.1 | 121 | 122.4 |
| $3–5,999 | 102 | 73.7 | 122 | 123.8 | 232 | 295.4 | 306 | 320.7 | 165 | 163.5 |
| $6–7,999 | 59 | 82.1 | 132 | 137.9 | 284 | 273.4 | 375 | 357.4 | 183 | 182.2 |
| $8–9,999 | 41 | 41.0 | 82 | 68.9 | 115 | 136.6 | 188 | 178.5 | 90 | 91.0 |
| $10–13,999 | 9 | 17.1 | 19 | 28.7 | 63 | 56.9 | 86 | 74.4 | 38 | 37.9 |

$\chi^2 = 40.916$

$\chi^2_{16}(.01) = 32.000$

Table 13. *Chi-Square Analysis* of Perceived Residential Quality of Region 5 by Income Category—Oak Hill Sample

| | Income Category | | | | | | | |
|---|---|---|---|---|---|---|---|---|
| | L.T. $3,000 | | $3–5,999 | | $6–7,999 | | G.T. $7,999 | |
| Residential Quality Response | Obs. | Ex. | Obs. | Ex. | Obs. | Ex. | Obs. | Ex. |
| 0 | 17 | 16.6 | 32 | 21.8 | 20 | 24.5 | 12 | 18.1 |
| 1 | 37 | 30.5 | 37 | 40.1 | 41 | 45.1 | 34 | 33.3 |
| 2 | 43 | 38.5 | 37 | 50.6 | 62 | 56.9 | 46 | 42.1 |
| 3 & 4 | 20 | 31.5 | 48 | 41.5 | 50 | 46.6 | 36 | 34.5 |

$\chi^2 = 20.204^*$

$\chi^2_9(.01) = 21.666$

* significant at 0.02 level

## FINAL COMMENTS

The process of action-space formation as discussed earlier is a complex and intriguing geographic phenomenon. This paper has analyzed several of the elements which hypothetically could influence the development and spatial parameters of the urban resident's action space. Tentatively, we may conclude that income plays a relatively minor role in the formation of the action space of individual urban residents; that is, action spaces are not a direct function of income, but rather are more dependent on the location of the neighborhood in which the city dweller lives. The action space of the individual is influenced not only by the location of his current residential location but also by the location of former residences. In general, lower-income groups view the quality of residential areas in a manner similar to that of middle groups, with the exception that they view residential areas as being of better quality than their middle-class suburban counterparts do.

The differences noted previously in this paper concerning the structure and extent of the action spaces of the populations in Cedar Hills and Oak Hill make it increasingly apparent that a single model of action space may not be appropriate. Rather, variations in the structure and morphology of the action space of central-city residents as opposed to suburban residents perhaps necessitate the construction of dual models. Given the diversity in the perception and structure of action spaces by the two groups, there still exists a communality of important urban features shared by the action spaces of both samples. The central business district, the major transportation link, and the major shopping center outside of the central business district in Cedar Rapids are all shared to some extent by the two sampled populations. Clearly, the second stage of the action-space formation process—community socialization—is a more formidable force in areas which contain relatively homogeneous physical structures. Cedar Hills, although displaying some income variation, appears to have a similarity of perspective which is absent in the Oak Hill area. Greater variation in the housing quality in Oak Hill may clearly contribute to this heterogeneity in Oak Hill in the sense that a broader range of life styles and residential preferences are found in the area.

The results given in this paper are a partial contribution to the continuing effort to model the action-space formation process.

# 6

# DIDO PATTERN ANALYSIS: GIGO OR PATTERN RECOGNITION?

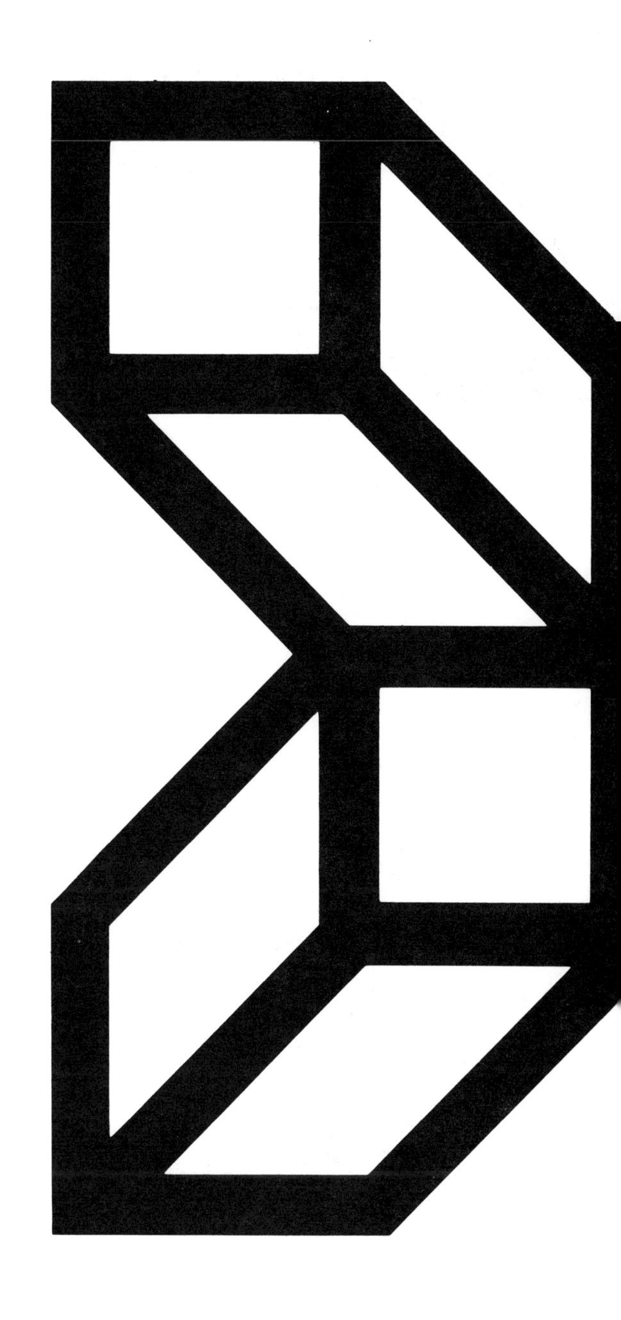

**Brian J. L. Berry**
The University of Chicago

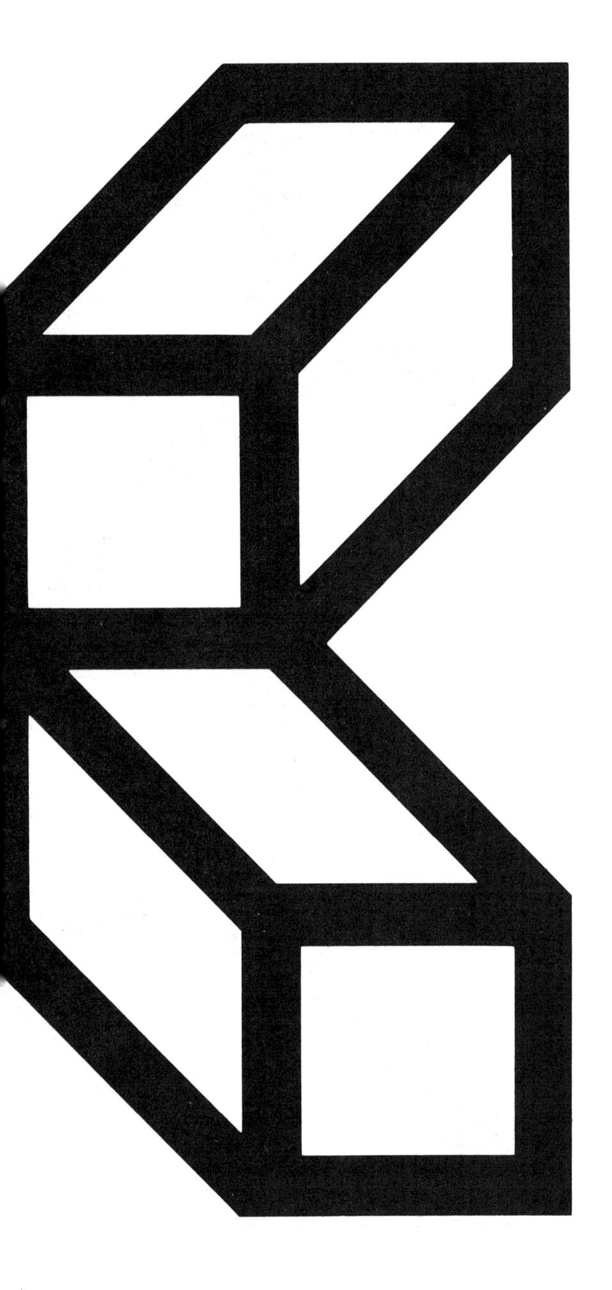

# DIDO Data Analysis: GIGO or Pattern Recognition?

**ABSTRACT**

Ecological factor analysis is characterized as DIDO data analysis (data in-dimensions out). Whether this analysis is a significant aid to pattern recognition, or a futile exercise in GIGO (garbage in-garbage out) depends upon a satisfactory resolution of a variety of questions about data, data manipulation, and the process of induction. These questions are reviewed by means of a presentation of three factor analyses of United States county data for 1950, 1960 and 1950–1960. The ultimate question of utility is answered by reverting to a phenomenological philosophy in which broader saliency is inferred through consistencies in analogous investigations.

The technical bases and attendant mathematical problems of ecological factor analysis as an approach to the study of multivariate spatial variation are increasingly well known (Janson 1969). Today, within geography, the majority of research workers who use factor analysis in their work most commonly have used (most frequently, *implicitly* must be added) the Hartshornian concept of areal differentiation as their working guide. Consistent with this framework, the primary task is assumed to be to describe how, and then to explain why, areas differ in the characteristics and behavior of their human populations (Rees 1970). I choose to call such an approach *DIDO* data analysis, because *data* are fed *into* a large computer with a packaged program in the hope that significant *dimensions* will come *out.* Advocates of this procedure claim that it provides an invaluable means of pattern recognition in mass data situations, when theory is relatively weak. Advocates of alternative research strategies charge that all too frequently bad data, inappropriate techniques, and unsound interpretation produce more *GIGO* than *DIDO. GIGO,* of course, means "garbage in—garbage out" and is a favorite acronym of the computist.

Undoubtedly, the charge and countercharge can probably never be satisfactorily resolved, for too many differences in personal values are involved. At least, it should be possible to isolate the most central issues involved, and thereby to bring at least a little order to the confrontation. This is the modest goal of this paper, which first proceeds step by step through an example of *DIDO* data analysis, raising a variety of the questions that arise in such analyses along the way, and then attempts to draw together the strengths and weaknesses of the procedures as a means of spatial pattern recognition.

## AN EXAMPLE: A FACTORIAL ECOLOGY OF THE UNITED STATES, 1950–1960

The example chosen is massive, because it is exactly in such situations that the need for *DIDO* data analysis might be thought to arise, and it is certainly in such circumstances that all the problems occur. As a starting point, I took the data available for each of the 3,102 countries of the United States in the 1950 and 1960 County and City Data Books and prepared three data matrices. Two matrices, of order 3,102 x 80, stored information on some 80 characteristics of each county in 1950 and 1960, respectively. A third matrix of the same order was prepared by calculating percentage changes in the value of each variable in each county in the 1950–60 decade. All three matrices were stored on high-density magnetic tape.

What do such matrices tell us about the county to county spatial variation in the United States in 1950 and

1960, and the 1950–60 decadal changes? Each one of us has a variety of notions, but geography simply does not have the theoretical capability today to propose a complete set of expectations. How might one then be able to begin to fill this gap? The factorial ecologist would respond by using data analysis designed specifically to determine (1) the number of types of spatial variation apparently present in the data; (2) the characteristics that exemplify each type of variation; and (3) the spatial pattern of the type. *DIDO* data analysis thus is brought into play. The steps are as follows:

(a) Descriptive statistics of each characteristic are calculated, and if the characteristic is not normally distributed, a transformation is found and applied to ensure normality.

(b) Each characteristic is then converted into a standard normal deviate, and the product-moment correlation of each characteristic with every other is computed.

(c) A choice is then made to perform either a principal components analysis, a principal axis (orthogonal) factor analysis, or an oblique factor analysis, on the correlation matrix. A decision is made about the number of "significant" factors, and perhaps an "analytic" axis rotation is performed.

(d) Matrices of factor loadings and factor scores are produced, the scores are mapped, and "interpretation" begins.

The questions that arise throughout these steps are obvious. What observations and which characteristics are to be used as a point of departure? Critics say that using whatever data may happen to be available smacks of opportunism and expediency and at best all that DIDO analysis can rediscover are the implicit concepts of the census taker; advocates say that prior determination is impossible, for the data themselves are the *terra incognita.* Why transform to normality? Critics say that results may be affected in unknown ways by this procedure; advocates say that it is simply a means of avoiding bias in calculating product-moment correlations. Why should one prefer one mode of data reduction as opposed to another? Critics say the choices made are too often determined by package-program availability; advocates may grudgingly agree, but say that principal components analysis simply provides an alternative set of axes to the original characteristics of the data, whereas factor analysis assumes that there is variance specific to each characteristic and that this specific variance should be eliminated from consideration in a study of covariant patterns. They go on to point out that oblique solutions demand even more specific knowledge about covariations at the onset. Thus, the sequence of factoring methods (principal components, principal axis factoring, oblique factor analysis) follows an equivalent knowl-

Table 1. *Size, Urban Density, Heterogeneity*

| Constituent Variables | Factor Loadings* 1960 Analysis | 1950 Analysis |
|---|---|---|
| Population | 0.968 | 0.971 |
| Number of families | 0.962 | 0.973 |
| Total deaths | 0.949 | 0.966 |
| Number of retail establishments | 0.946 | 0.956 |
| Population in kindergarten through high school | 0.946 | 0.965 |
| Total labor force | 0.942 | 0.975 |
| Total births | 0.941 | 0.963 |
| Number of dwelling units | 0.931 | 0.973 |
| Number of employees, sales | 0.929 | 0.958 |
| Number of employees, transportation | 0.914 | 0.921 |
| Rank | −0.907 | −0.937 |
| Number of employees, miscellaneous | 0.897 | 0.928 |
| Retail sales in $1,000 | 0.883 | 0.922 |
| Retail payroll in $1,000 | 0.882 | 0.927 |
| Marriages | 0.851 | 0.754 |
| Number of manufacturing establishments | 0.844 | 0.937 |
| Number of wholesale establishments | 0.844 | 0.793 |
| Deposits | 0.842 | 0.739 |
| Wholesale payroll in $1,000 | 0.815 | 0.829 |
| Wholesale sales | 0.796 | 0.677 |
| Density | 0.789 | 0.798 |
| Savings | 0.761 | 0.735 |
| Number of employees, manufacturing | 0.728 | 0.922 |
| Manufacturing payroll in $1,000 | 0.657 | 0.835 |
| Number of manu. estab. 20–99 employees | 0.655 | 0.850 |
| Value added in $1,000 | 0.654 | 0.822 |
| Number of manu. estab. more than 99 employees | 0.513 | 0.808 |
| Percent families, $3,000 income and under 1960 | −0.419 | ... |
| Percent families, $10,000 income and over 1960 | 0.427 | ... |
| Percent increase population | 0.412 | 0.482 |
| Percent labor force male | −0.493 | ... |
| Migration (different residences, 1950) | ... | 0.896 |
| Percent of population urban | ... | 0.703 |
| Value of dairy products | ... | 0.509 |
| Number of farms | ... | 0.421 |
| Value of poultry products | ... | 0.455 |
| Percent of housing units with sound plumbing | ... | 0.444 |

* Note: All variables with factor loadings lying in the range +0.4 to −0.4 are omitted.

edge sequence. Why rotate factorial results? Critics say that axis rotation confounds mathematical solutions; advocates, that it aids in interpretation.

There are other questions too, but those outlined above should serve to give some flavor of the debate, although not of its animosities and mutual charges of incompetence. Thus, notwithstanding the obvious criticism that could be forthcoming, in developing the examples that follow I adopted my standard procedure of taking the most simple-minded options in each case. After transformation and computation of the correlation coefficients, principal components analysis was applied. Then, accepting the argument advanced by some that no component with variance less than unity (the variance of an original variable) can possibly be meaningful, only those components with eigenvalues (variance of components) exceeding unity were rotated, using Kaiser's "normal varimax" criterion, to a position approximating "simple structure." This was in the attempt to ensure that each characteristic as nearly as possible correlated highly with only one component. As a result, the factor loadings (correlations of characteristics with components) could be used to divide the characteristics into sub-sets, each displaying a particular type of spatial variation, and the factor scores (scores of observations on components) could be used when mapped to display the spatial variation of the type. The combination of the cluster of variables and the map of factor scores became the basic ingredient in interpreting the dimensions of variation emerging at the end of the process.

## Structure, 1950 and 1960

The first two analyses involved identical studies of 3,102 x 80 data matrices using information reported in the 1950 and 1960 County and City Data books. All definitions thus follow standard United States census usage. Results are portrayed in tables 1–7. First, the two analyses produced five near-identical primary factors, labeled "Size, Urban Density, Heterogeneity" in the spirit of Louis Wirth (table 1), "Agricultural Intensity" (table 2), "Socio-economic Status" (table 3), "Stage in Life Cycle" or "Family Structure" (table 4), and "National Market Access and Density" (table 5). There were four lesser factors in 1960 (table 6) and three in 1950 (table 5).

Examination of the variables that cluster in each case leaves no doubt about the bases of the interpretation, and we conclude that, in so far as the County and City Data books describe characteristics that Americans consider it important to have sampled and reported, there are five principal themes demanding consideration. Figures 1–5 map the factor scores to highlight the accompanying spatial patterns. Figure 1 clearly brings out the nation's metropolitan structure; the areas identified concentrate the nation's population, housing, economic activity, and gross assets. Figure 2 reveals those areas oriented most to agricultural pursuits. Figure 3 picks out high- and low-status regions. Of interest is the fact that the South, other peripheral areas in which minority groups live (American Indians, Spanish Americans), and the cores of the largest cities share the same low status today. Figure 4 picks out the areas of aging, declining populations and those with particularly youthful demographic characteristics. Figure 5 reveals how farm sizes increase and densities drop systematically with increasing distance from the nation's heartland.

At this point, however, the first problem of this form of analysis emerges. We called the first five factors "primary" because they accounted for the greatest part of the total variance and because they were repeated in the two analyses. Yet this judgment is clearly determined by the data inserted into the analysis; the more variables of a particular kind, the more likely is one to derive a primary dimension out of the analysis. Equally interesting spatial patterns characterize lesser factors. Take, for example, the first and third of the residual factors in table 6. When their scores are mapped, figures 6 and 7 result. Figure 6 outlines the nation's problem regions of 1960; many of them are now covered by Regional Development Commissions. Figure 7 picks out the facet of American industrial life most forgotten—the widespread diffusion into small towns beyond the traditional manufacturing belt of labor-intensive manufacturing employing the lowest proportions of males. Who is to say that these last two spatial patterns are any less fundamental than the first five in developing a broad-scale understanding of the contemporary American landscape? Criteria external to DIDO data analysis are not only important in making such an assessment; they are essential. Of itself, the data analysis provides no guidance and may, through focusing attention on eigenvalues that are an artifactual consequence of the data input, be downright misleading.

Table 2. *Agricultural Intensity*

| Constituent Variables | Factor Loadings* 1960 Analysis | 1950 Analysis |
|---|---|---|
| Number of commercial farms | 0.891 | 0.865 |
| Value of farm products | 0.883 | 0.927 |
| Number of farms | 0.798 | 0.705 |
| Value livestock | 0.774 | 0.846 |
| Value all crops | 0.738 | 0.795 |
| Acres in farmland | 0.666 | 0.826 |
| Percent land in farms | 0.617 | ... |
| Value poultry | 0.521 | 0.551 |
| Value dairy | 0.479 | 0.509 |
| Percent operated by tenants | 0.522 | 0.600 |
| Average farm value | ... | 0.683 |
| Average farm size | ... | 0.419 |

* See Note to Table 1.

Table 3. *Socio-Economic Status*

| Constituent Variables | Factor Loadings* 1960 Analysis | 1950 Analysis |
|---|---|---|
| Percent of population with Grade 12 education | 0.875 | 0.790 |
| Percent of population with Grade 5 education | −0.842 | −0.751 |
| Median family income | 0.816 | ... |
| Percent families < $3,000—1960; < $2,000—1950 | −0.803 | −0.840 |
| Median school years completed | 0.783 | 0.505 |
| Percent units with sound plumbing | 0.778 | 0.732 |
| Percent families > $10,000—1960; > $5,000—1950 | 0.749 | 0.825 |
| Average farm value | 0.599 | ... |
| Percent nonwhite | −0.527 | −0.459 |
| Percent population employed | 0.499 | 0.687 |
| Percent increase population | 0.457 | 0.415 |
| Percent labor force male | ... | 0.513 |

* See Note to Table 1.

Table 4. *Stage in Life Cycle* (Family Structure)

| Constituent Variables | Factor Loadings* 1960 Analysis | 1950 Analysis |
|---|---|---|
| Percent population under 5 | 0.890 | 0.673 |
| Percent population over 21 | −0.889 | −0.660 |
| Percent population over 65 | −0.880 | −0.869 |
| Percent increase population | 0.431 | ... |
| Median age | ... | −0.843 |

* See Note to Table 1.

Table 5. *National Market Access and Density*

| Constituent Variables | Factor Loadings* 1960 Analysis | 1950 Analysis |
|---|---|---|
| Population potential | −0.668 | −0.664 |
| EMT | 0.607 | 0.697 |
| Average farm size | 0.589 | 0.542 |
| Acres in farmland | 0.447 | ... |
| Density | ... | −0.400 |
| Percentage unemployed | ... | 0.606 |

* See Note to Table 1. EMT is estimated total travel distance to other parts of the nation.

Table 6. *Other 1960 Factors*

| Constituent Variables | Factor Loadings |
|---|---|
| Percent operated by tenants | 0.604 |
| Percent unemployed | −0.564 |
| Percent land in farms | 0.491 |
| Average farm value | 0.487 |
| | |
| Manufacturing payroll in $1,000 | −0.529 |
| Value added in $1,000 | −0.524 |
| Percent population employed | −0.480 |
| Number of employees, manufacturing | −0.470 |
| | |
| Percent labor force male | 0.549 |
| Number of employees, manufacturing | −0.482 |
| | |
| Net migration | 0.782 |
| Percent units owner-occupied | 0.589 |

Table 7. *Other 1950 Factors*

| Constituent Variables | Factor Loadings |
|---|---|
| Percent units owner-occupied, 1950 | −0.785 |
| Percent nonwhite, 1950 | −0.548 |
| Value dairy, 1952 | −0.443 |
| Value poultry, 1952 | −0.405 |
| | |
| Median family income | −0.539 |
| Median school years completed | −0.459 |
| | |
| Percent land in farms | 0.957 |

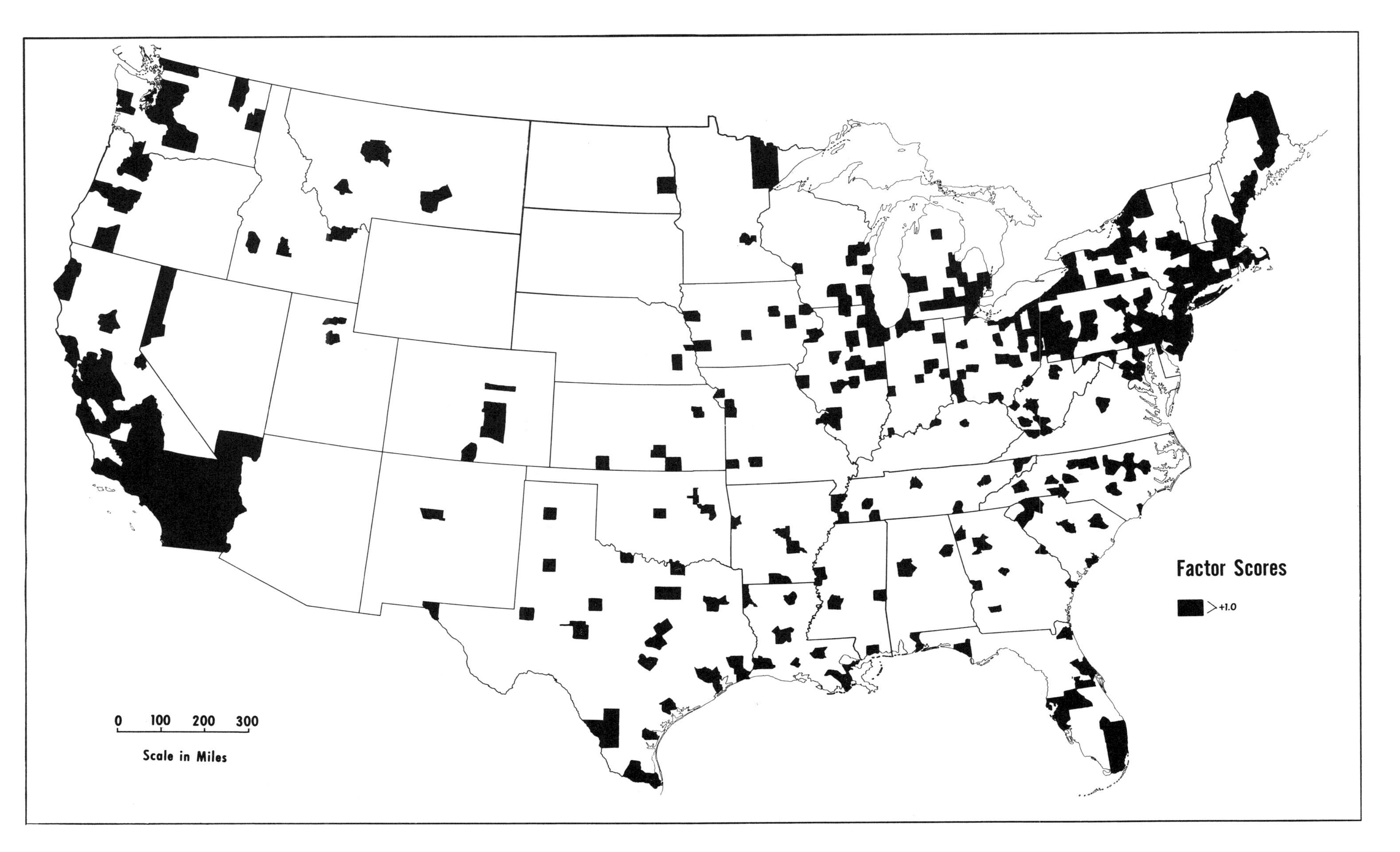


Figure 1. URBAN SIZE, DENSITY AND HETEROGENEITY, 1960

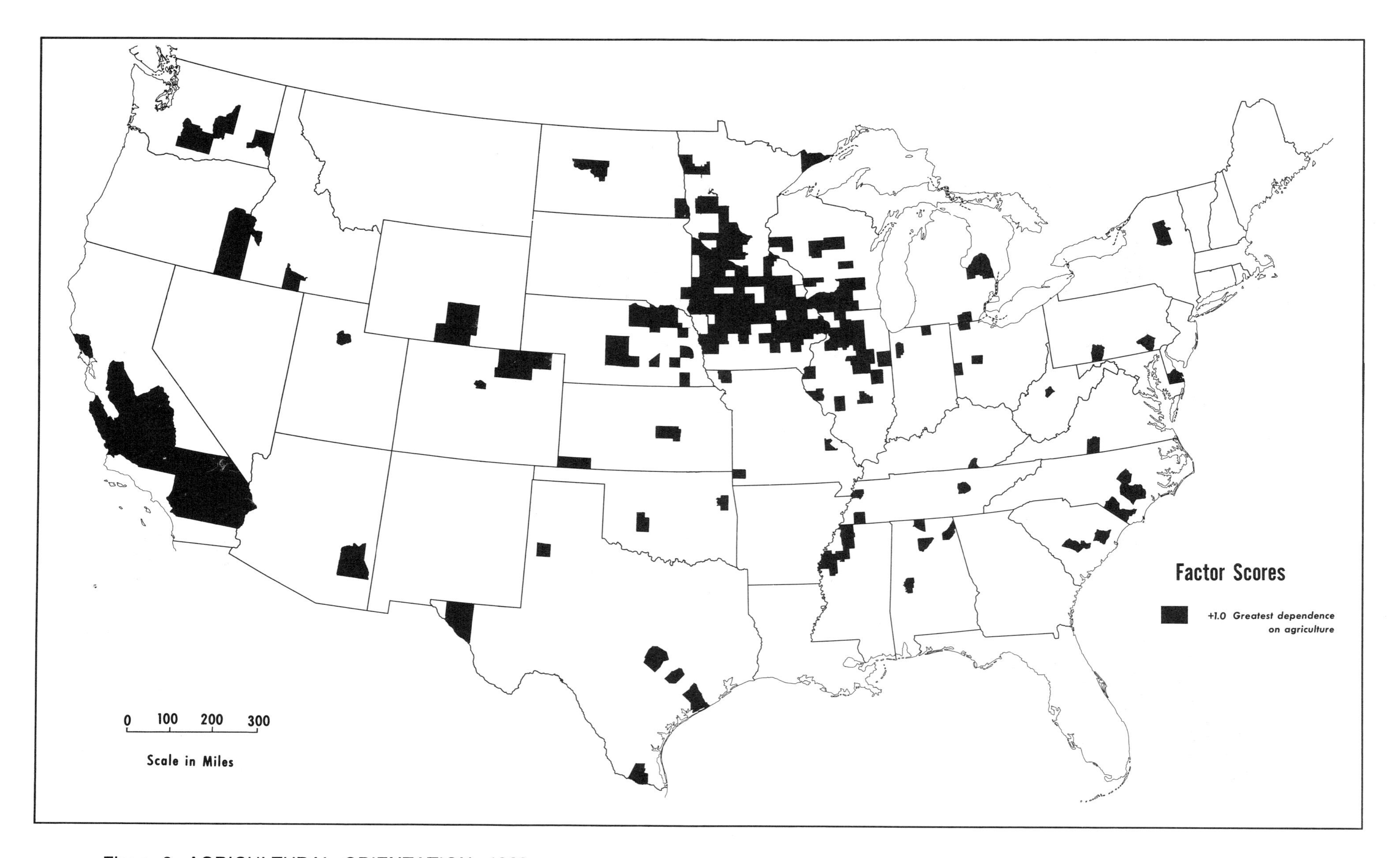


Figure 2. AGRICULTURAL ORIENTATION, 1960

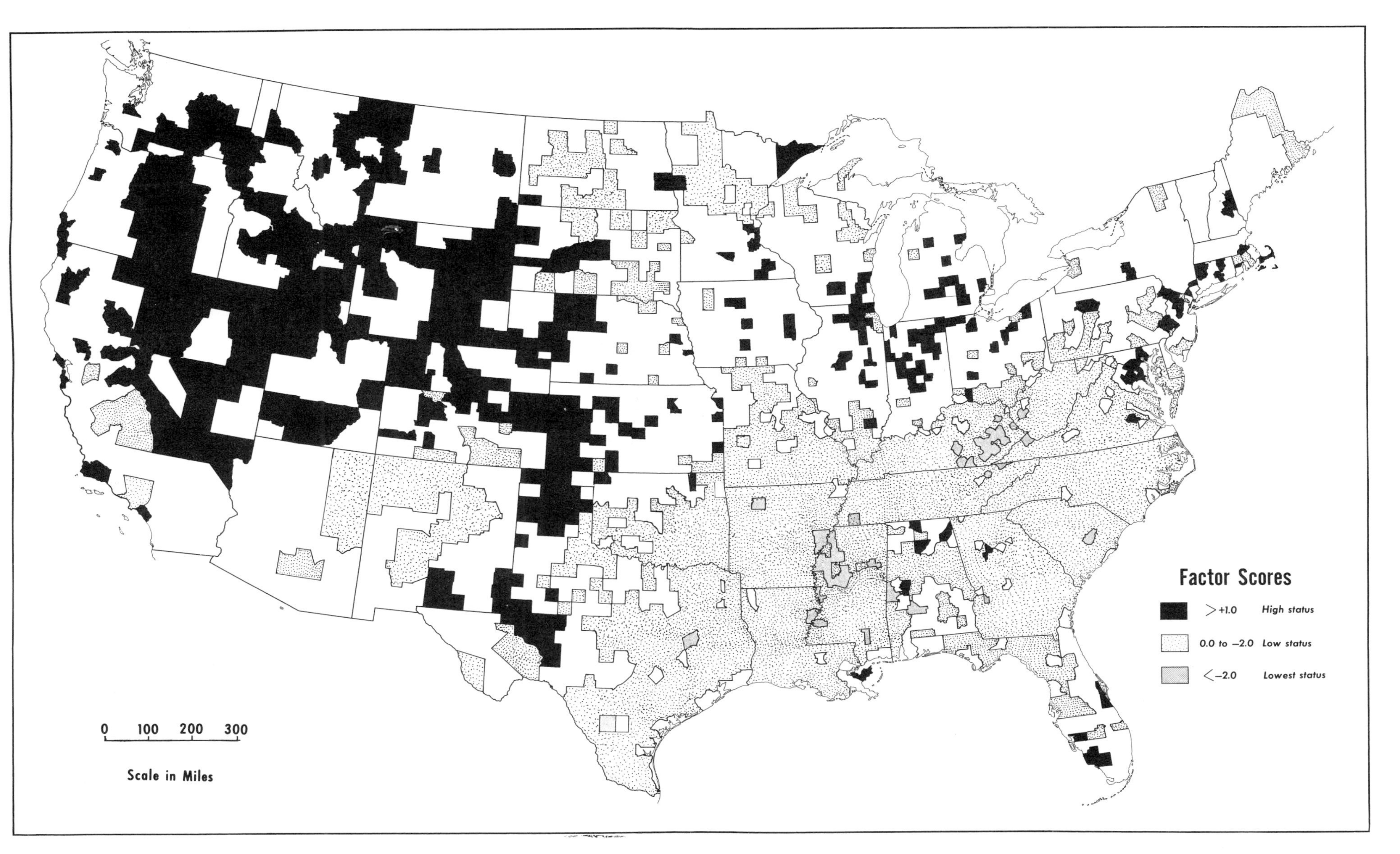


Figure 3. SOCIO-ECONOMIC STATUS, 1960

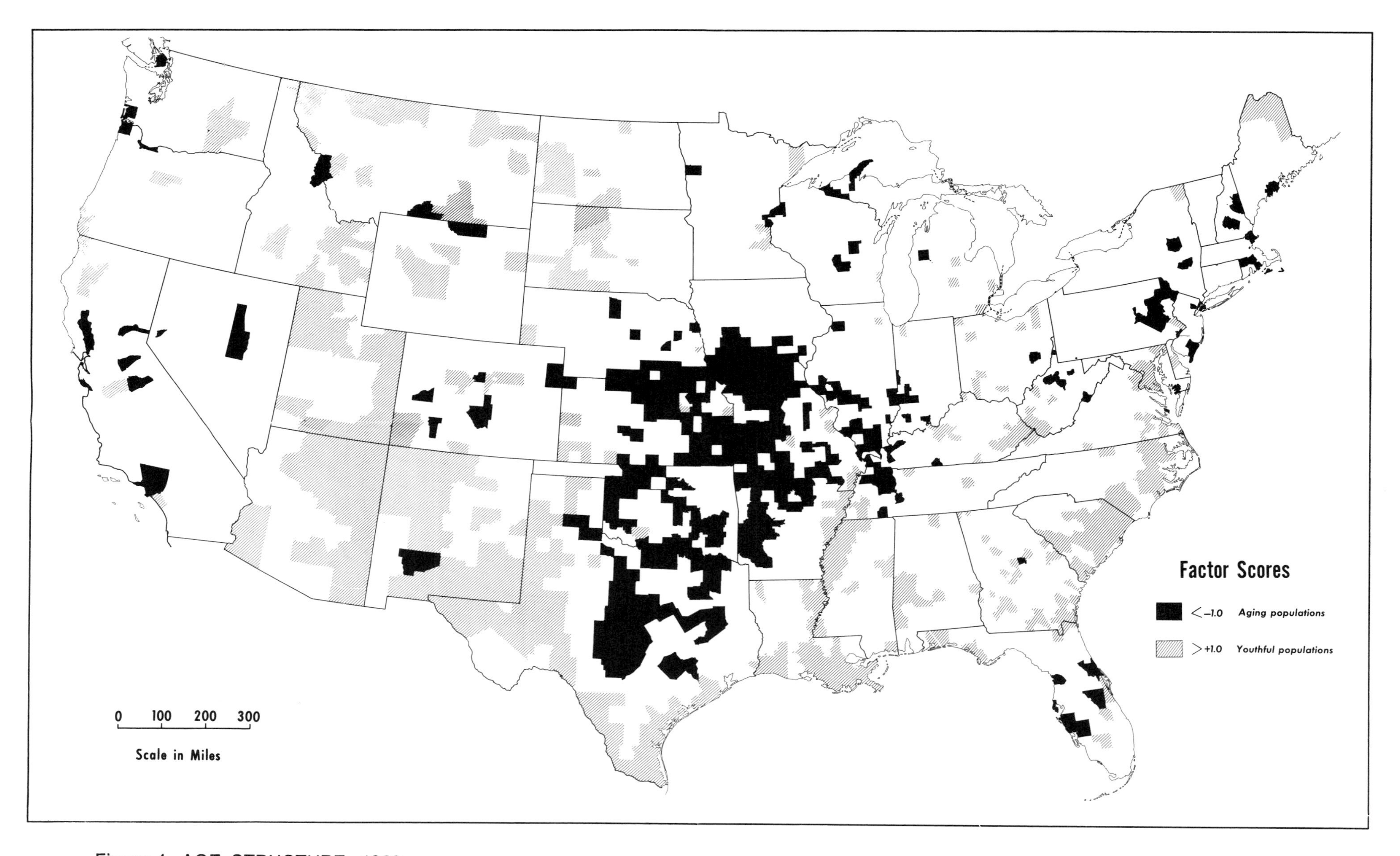


Figure 4. AGE STRUCTURE, 1960

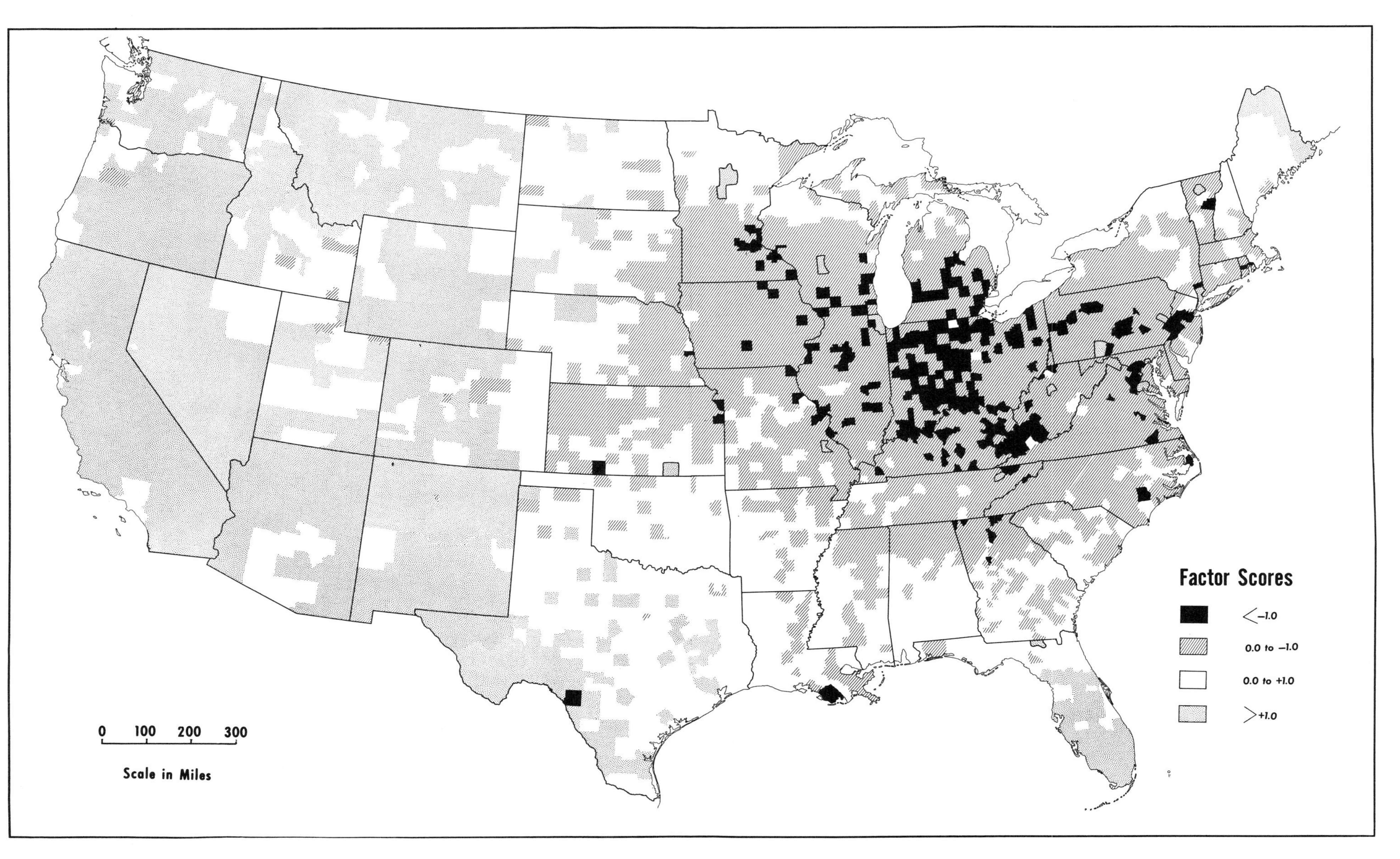


Figure 5. ACCESSIBILITY AND FARM SIZE, 1960

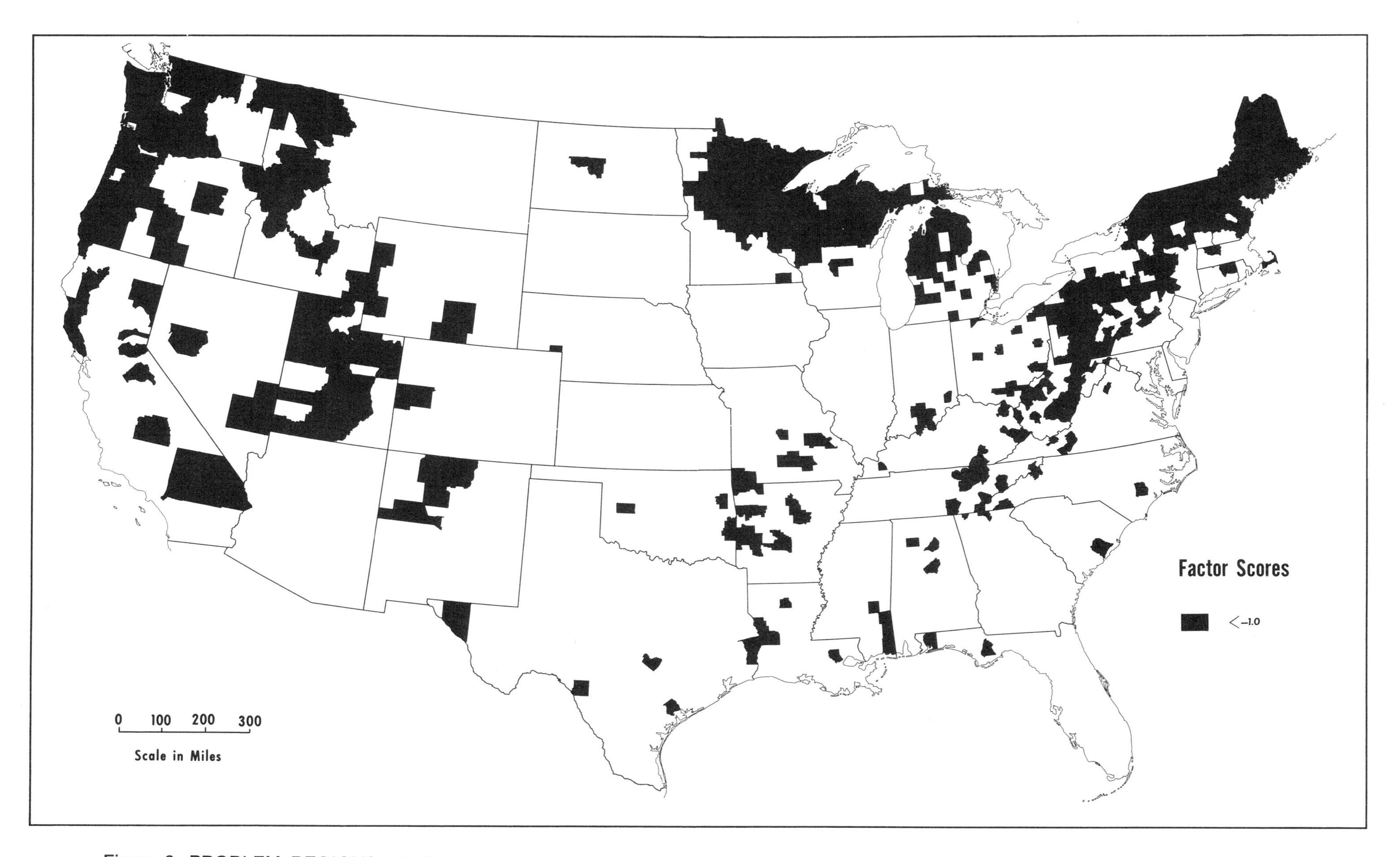


Figure 6. PROBLEM REGIONS, 1960

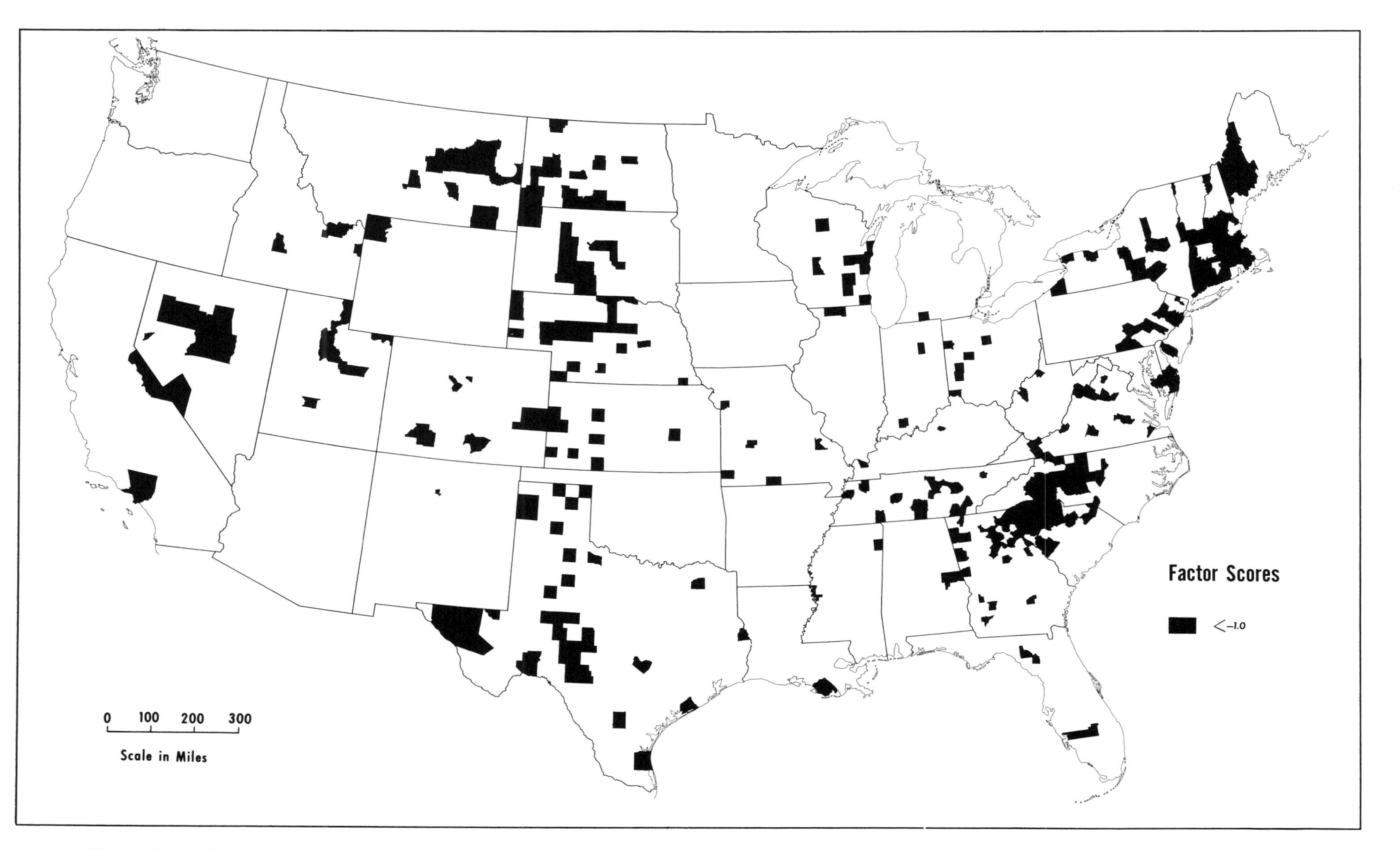


Figure 7. SMALL TOWN MANUFACTURING ORIENTATION, 1960

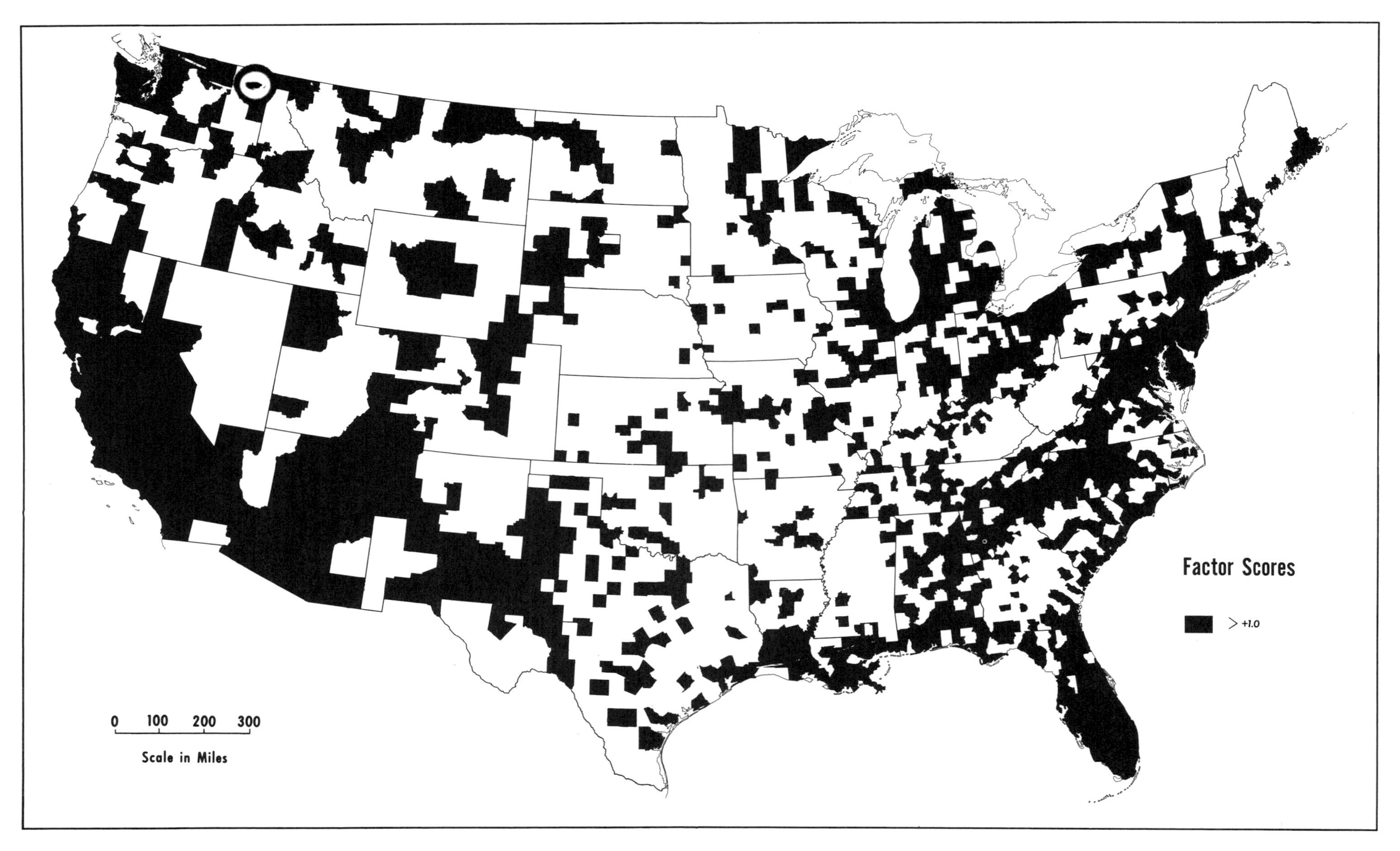


Figure 8. AREAS OF GREATEST PROPORTIONATE GROWTH OF DENSITY AND LABOR FORCE, 1950–1960

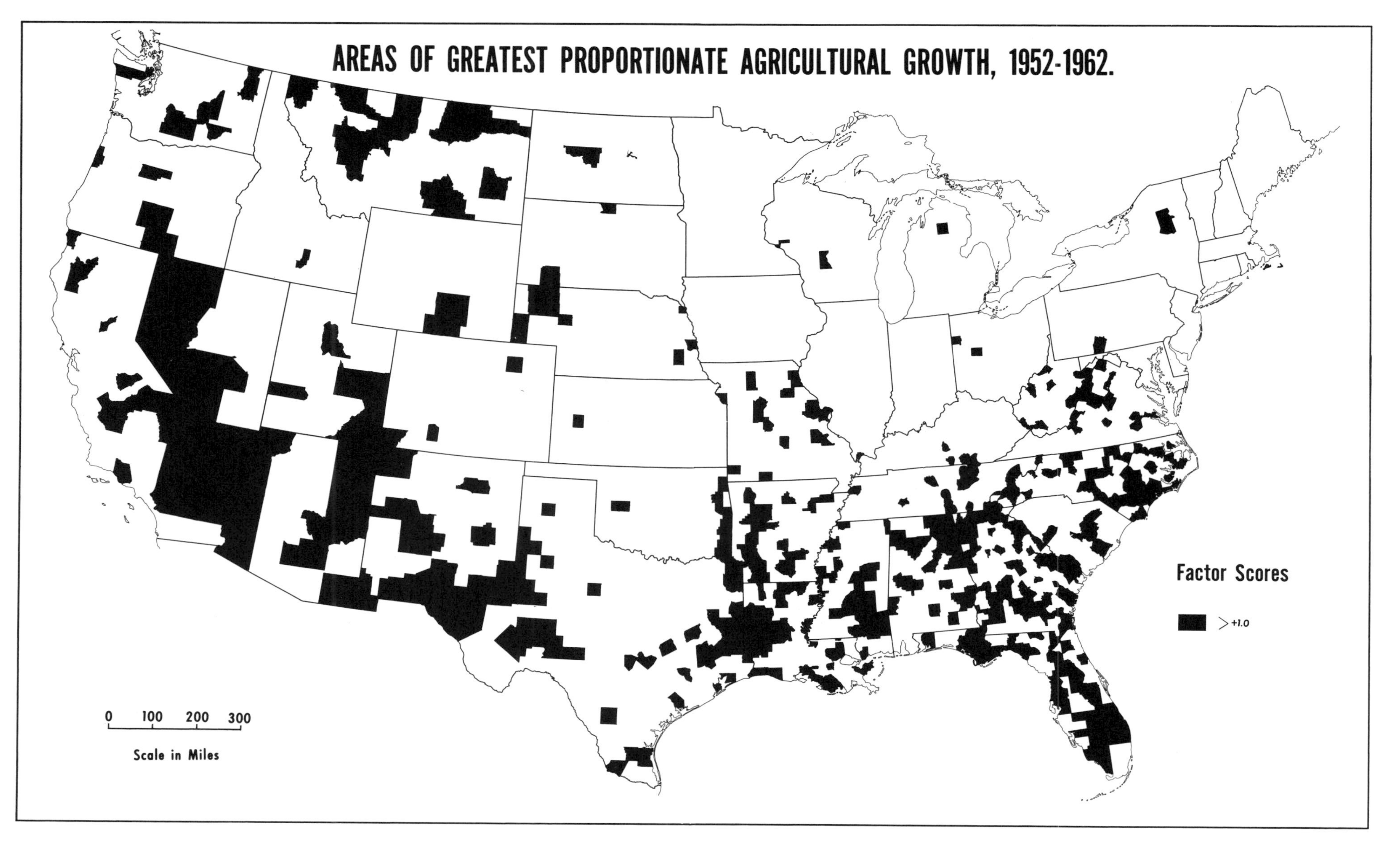


Figure 9. AREAS OF GREATEST PROPORTIONATE AGRICULTURAL GROWTH, 1952–1962

Table 8. *Increasing Size and Density, 1950–1960*

| Constituent Variables | Factor Loadings |
|---|---|
| Labor force 60/50 | 0.978 |
| Number of retail establishments 62/52 | 0.974 |
| Deaths 62/52 | 0.973 |
| Density 60/50 | 0.970 |
| Births 62/52 | 0.955 |
| Number of families 60/50 | 0.689 |
| Percent nonwhite 60/50 | 0.538 |
| Kindergarten through high school 60/50 | 0.574 |
| Number of employees, miscellaneous 60/50 | 0.476 |
| Retail sales in $1,000 62/52 | 0.431 |
| Number of dwelling units | 0.425 |
| Deposits 62/52 | 0.410 |

Table 9. *Agricultural Change, 1952–1962*

| Constituent Variables | Factor Loadings |
|---|---|
| Value of farm products 62/52 | 0.834 |
| Number of commercial farms 62/52 | 0.799 |
| Average farm size 62/52 | 0.756 |
| Number of farms 62/52 | 0.750 |
| Percent operated by tenants 62/52 | 0.729 |
| Value livestock 62/52 | 0.724 |
| Value dairy 62/52 | 0.697 |
| Value of all crops 62/52 | 0.680 |
| Value poultry 62/52 | 0.626 |
| Acres in farmland 62/52 | 0.561 |
| Retail sales in $1,000 62/52 | 0.561 |

## Change 1950–1960

When similar analysis was performed on the 3,102 x 80 change matrix, a more complex ten-factor structure emerged. Tables 8–17 display the interpreted results. Separate dimensions apparently differentiate (1) increasing size and density (table 8); (2) increasing agricultural intensity (table 9); (3) a combination of suburban and outer-rim growth (table 10); (4) increased aging of the population (table 11); (5) increasing numbers of manufacturing establishments (table 12); (6) change of status (table 13); (7) zones of population and wholesaling growth (table 14); (8) change in manufacturing and value added (table 15); (9) areas of increasing disadvantaged population (table 16); and (10) zones of increasing manufacturing employment (table 17).

Recalling that the data are based upon percentage change, so that growth from a small base can result in a large percentage, figures 8–16 map the resulting factor scores. A progressively greater scale of activity and increasing density was characteristic of the metropolitan zones and the "outer rim" of the West and South (figure 8). The South and West enjoyed the greatest agricultural growth in the decade (figure 9). Population and dwelling units increased most rapidly in outer suburbia and in the southern and western rims (figure 10). On the other hand, the nation's population in the outer peripheries aged progressively (figure 11). Many of the nation's central cities declined in relative status, while some increases were registered in zones previously relatively low on the nation's social ladder (figure 12). Wholesaling expanded along with growing populations in the West (figure 13). Manufacturing payrolls and value added showed the greatest proportional growth in the outer rim (figure 14). Some regions declined (figure 15) and upward percentage shifts in manufacturing were registered in many areas outside the traditional manufacturing belt (figure 16).

Table 10. *Suburban and Outer Rim Growth, 1950–1960*

| Constituent Variables | Factor Loadings |
|---|---|
| Number of employees, sales 60/50 | 0.872 |
| Number of employees, transportation | 0.735 |
| Number of dwelling units 60/50 | 0.717 |
| Number of employees, miscellaneous 60/50 | 0.674 |
| Kindergarten through high school 60/50 | 0.593 |
| Percent units owner occupied 60/50 | 0.522 |
| Number of families 60/50 | 0.514 |
| Percent population under 5 60/50 | 0.502 |
| Percent population employed 60/50 | −0.408 |

Table 11. *Aging of the Population, 1950–1960*

| Constituent Variables | Factor Loadings |
|---|---|
| Percent population over 65 60/50 | 0.727 |
| Percent population over 21 60/50 | 0.595 |
| Population 60/50 | −0.507 |
| Percent population employed 60/50 | 0.465 |
| Median age 60/50 | 0.441 |

Table 12. *Manufacturing Change, 1952–1962 (1),* Numbers and Size

| Constituent Variables | Factor Loadings |
|---|---|
| Number of manu. estab. 20–99 employees 62/52 | 0.701 |
| Number of manu. estab. 99+ employees 62/52 | 0.682 |
| Number of manufacturing establishments 62/52 | 0.594 |
| Value added in $1,000 62/52 | 0.591 |

Table 13. *Change of Status, 1950–1960*

| Constituent Variables | Factor Loadings |
|---|---|
| Median school year 60/50 | 0.871 |
| Median family income 60/50 | 0.789 |
| Marriages 62/52 | 0.626 |

Table 14. *Population Increase and Change in Wholesaling, 1950–1960*

| Constituent Variables | Factor Loadings |
|---|---|
| Number of wholesale establishments 62/52 | 0.606 |
| Wholesale sales 62/52 | 0.558 |
| Population 60/50 | 0.550 |
| Average farm value 62/52 | 0.503 |

Table 15. *Manufacturing Change, 1952–1962 (2),* Payroll and Value Added

| Constituent Variables | Factor Loadings |
|---|---|
| Number of employees, manufacturing 62/52 | 0.690 |
| Manufacturing payroll in $1,000 62/52 | 0.660 |
| Value added in $1,000 62/52 | 0.521 |

Table 16. *Change among the Disadvantaged, 1950–1960*

| Constituent Variables | Factor Loadings |
|---|---|
| Percent units with sound plumbing 60/50 | −0.773 |
| Percent unemployed 60/50 | −0.619 |
| Grade 12 60/50 | −0.475 |

Table 17. *Manufacturing Employment Growth, 1950–1960*

| Constituent Variables | Factor Loadings |
|---|---|
| Number of employees, manufacturing 60/50 | −0.597 |
| Percent units owner-occupied 60/50 | 0.554 |

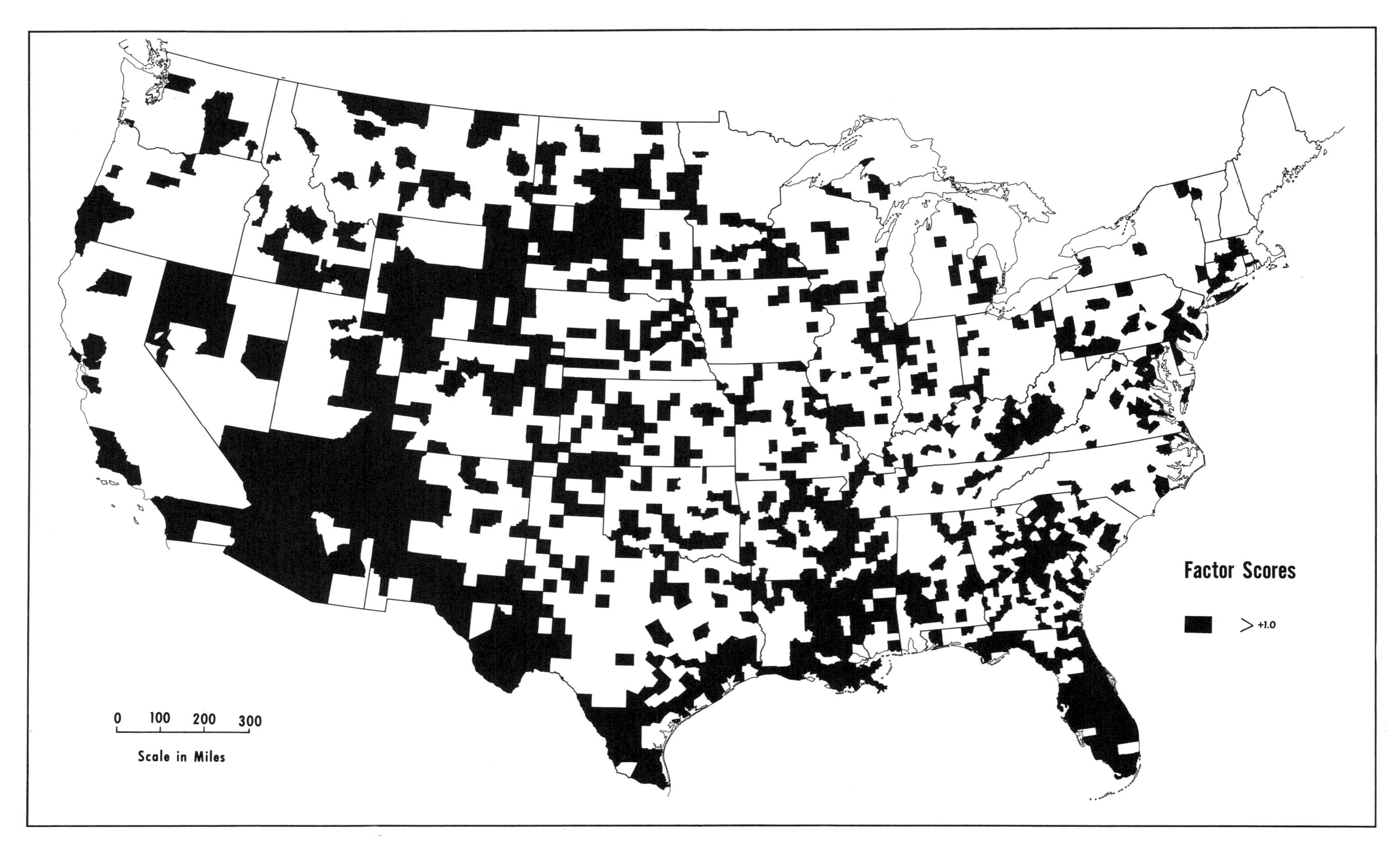


Figure 10. AREAS OF GREATEST PROPORTIONATE GROWTH IN POPULATION AND DWELLING UNITS, 1950–1960

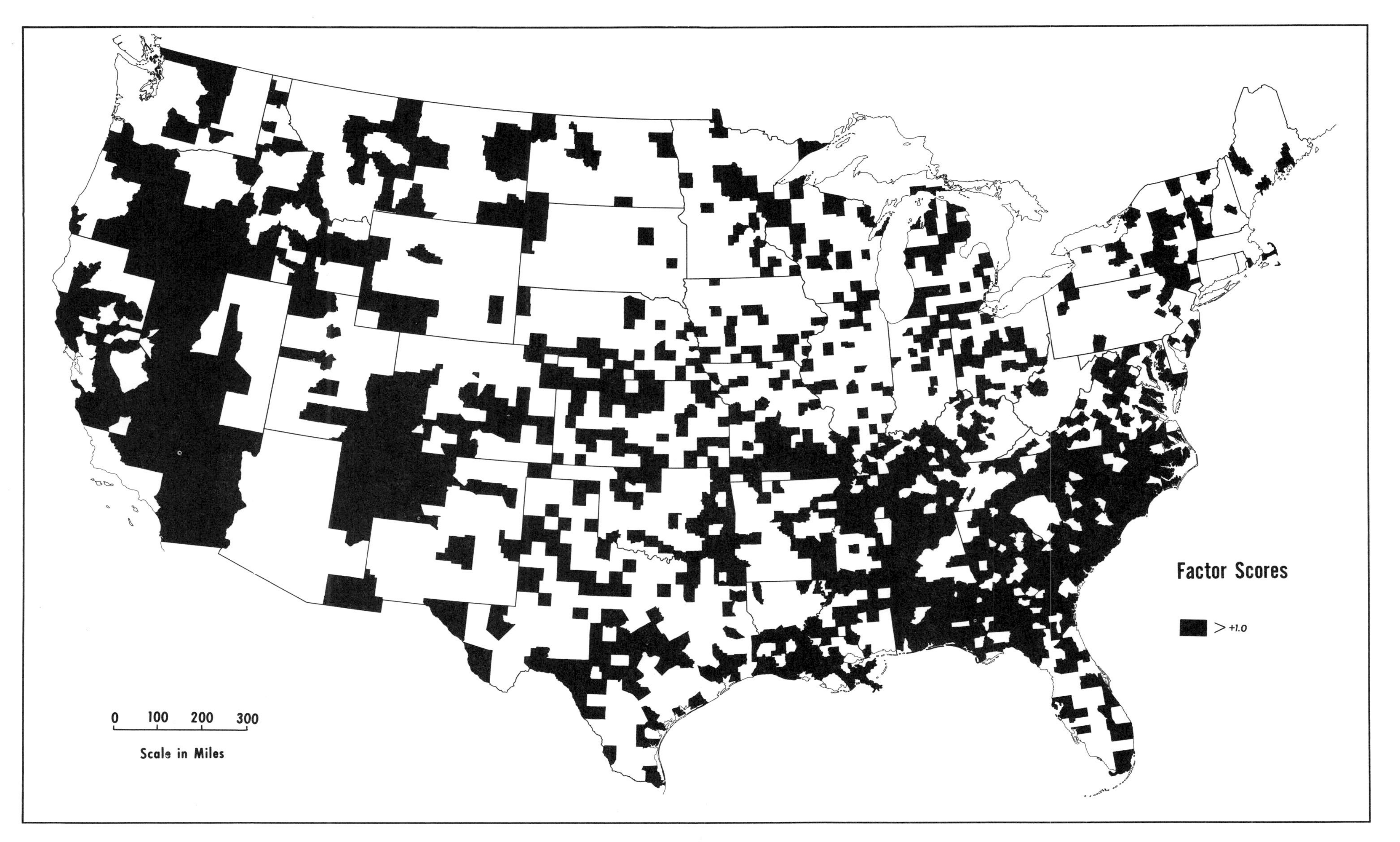


Figure 11. AREAS OF GREATEST PROPORTIONATE AGING OF THE POPULATION, 1950–1960

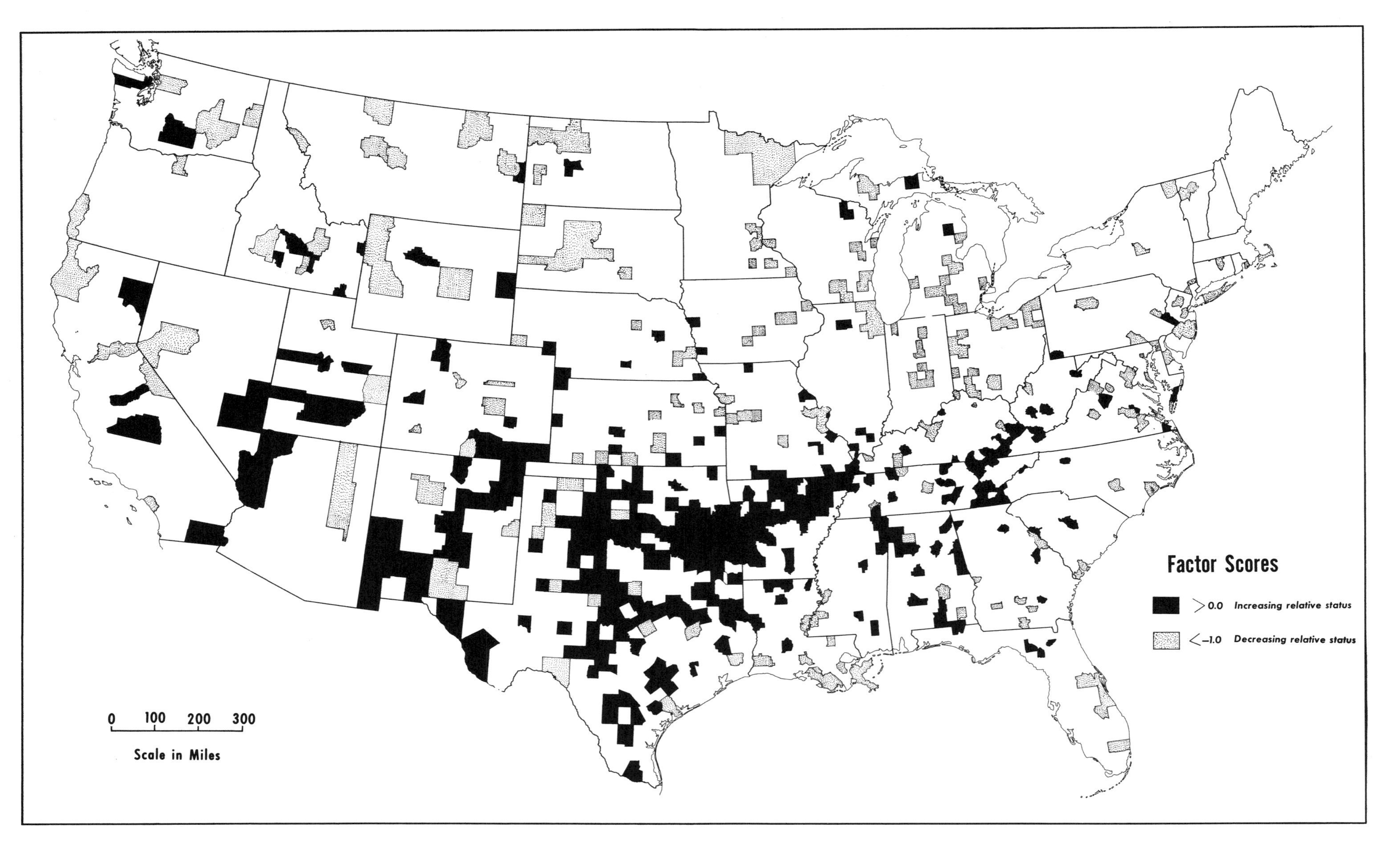


Figure 12. AREAS OF GREATEST RELATIVE STATUS CHANGE, 1950–1960

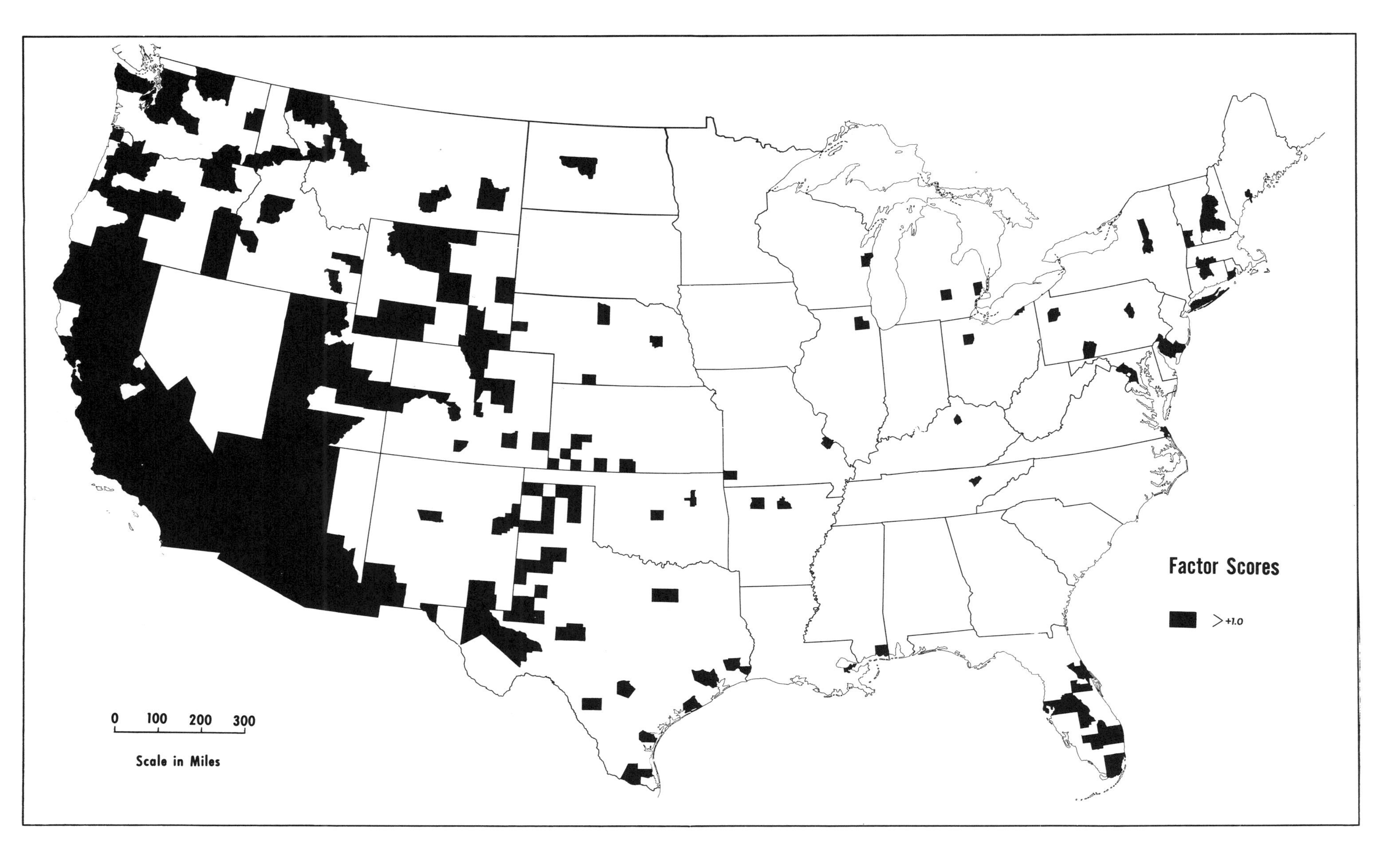


Figure 13. AREAS OF GREATEST GROWTH IN WHOLESALING ACTIVITIES, 1952–1962

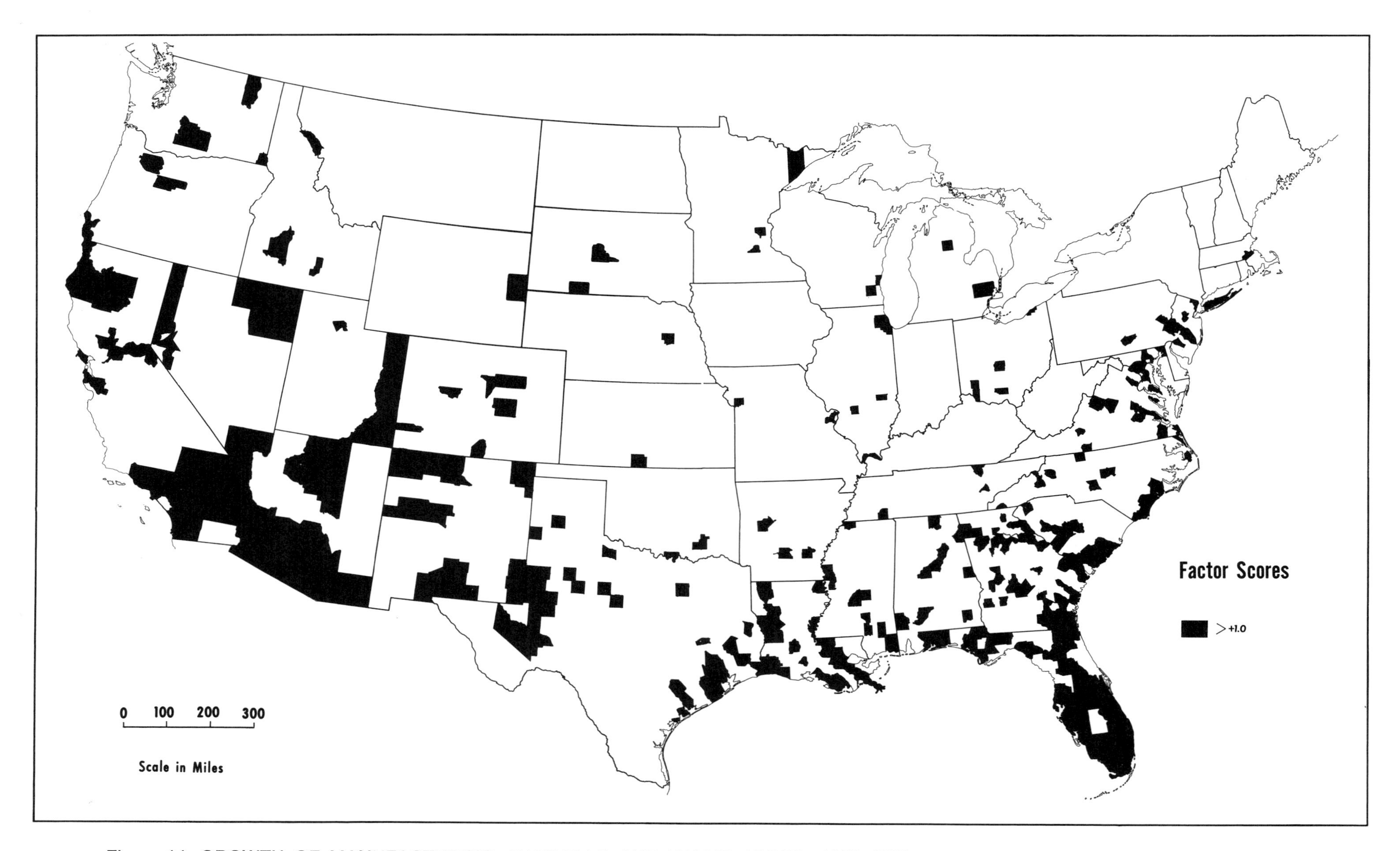


Figure 14. GROWTH OF MANUFACTURING, PAYROLLS AND VALUE ADDED, 1952–1962

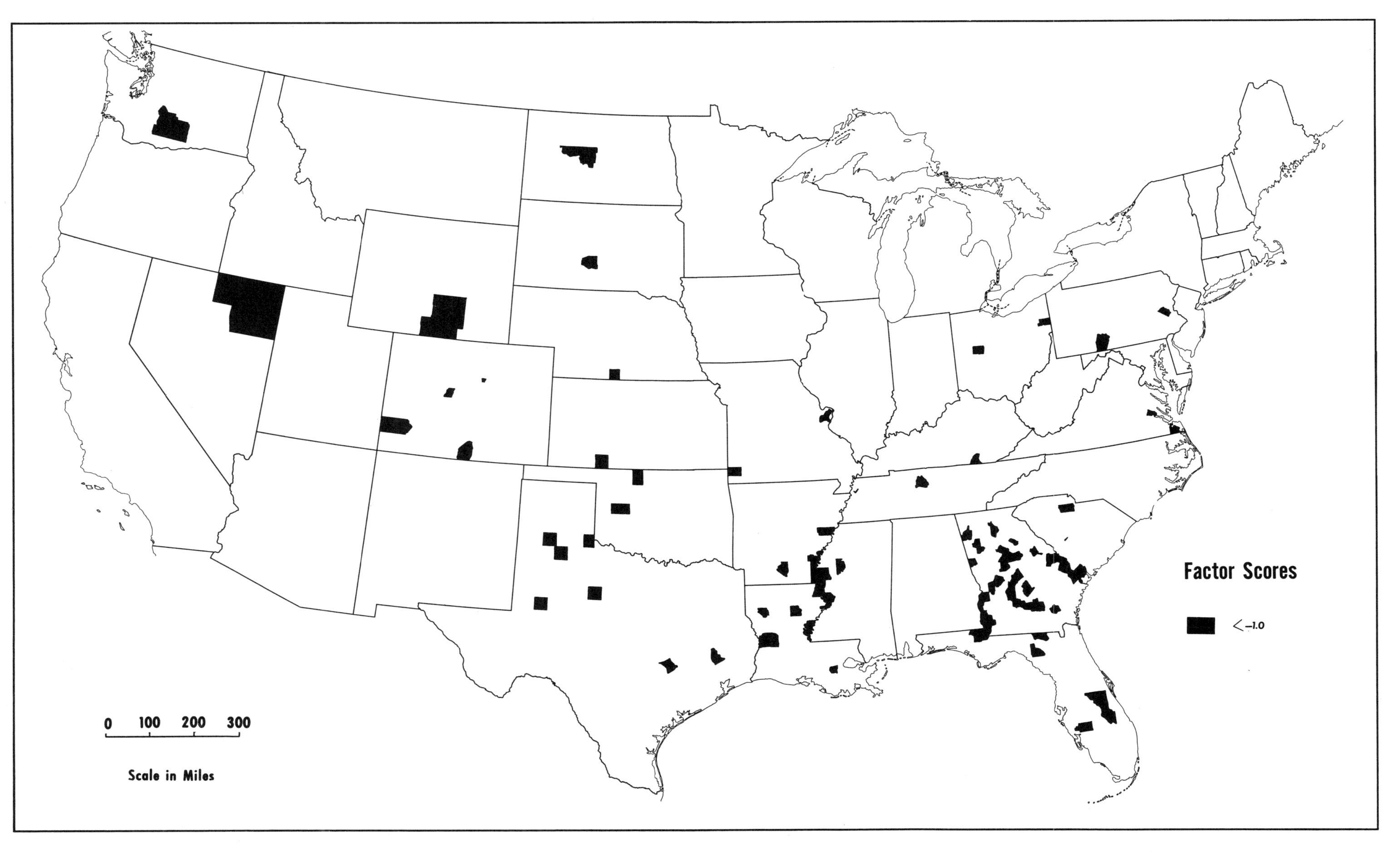


Figure 15. AREAS OF GREATEST RELATIVE INCREASE IN DISADVANTAGED POPULATION, 1950–1960

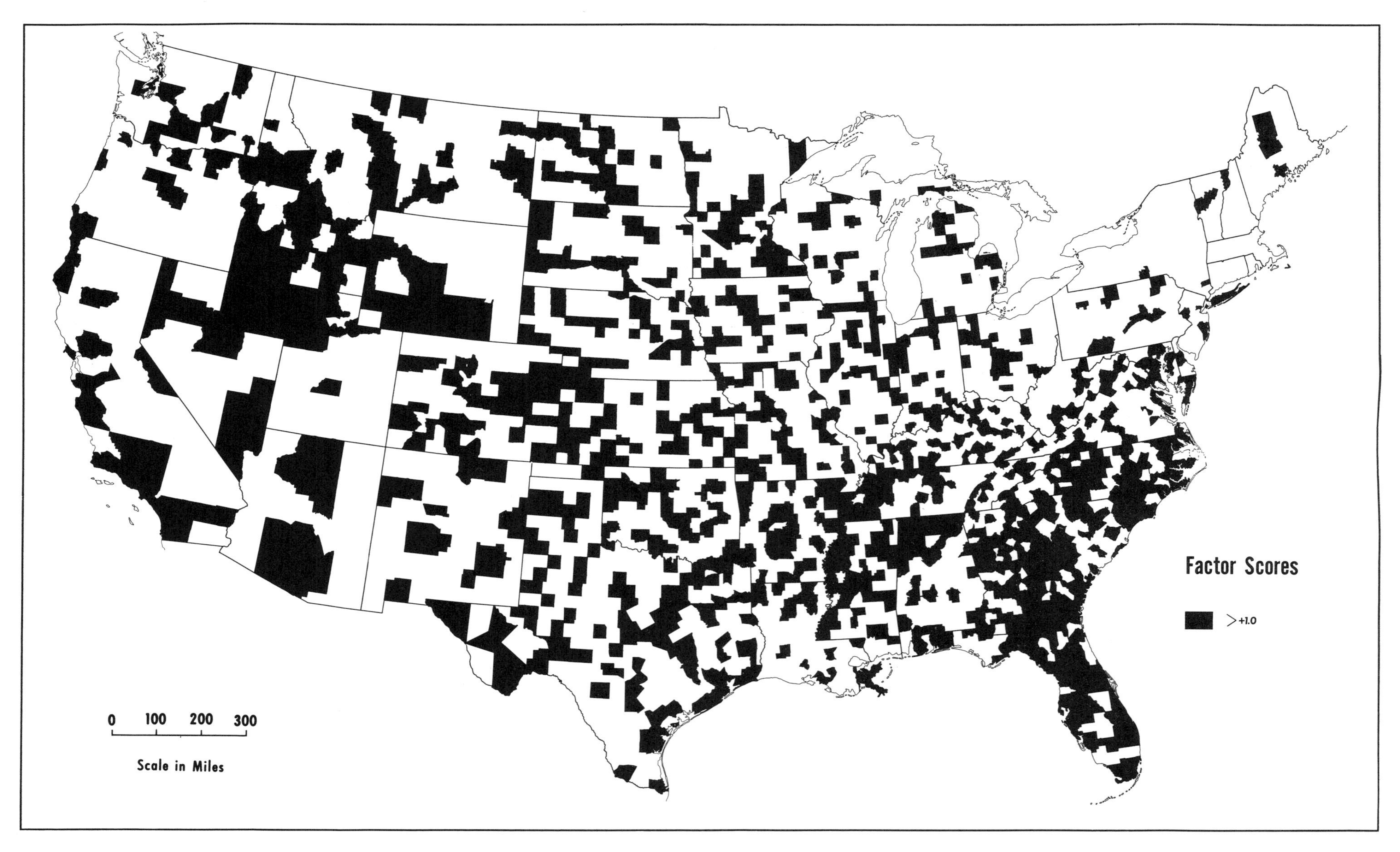


Figure 16. AREAS OF GREATEST PROPORTIONATE EXPANSION OF MANUFACTURING EMPLOYMENT, 1952–1962

## HAS ANYTHING WORTHWHILE RESULTED?

If nothing else, the DIDO data analysis procedures have produced a variety of maps and, if one were to revert to the traditional cartographic rationale of useful results, such a resulting set of patterns of covariance would be evidence enough of utility. Do the patterns have some broader saliency? They are consistent with results of similar exercises performed with data for United States cities (Berry 1970b) and with results of analogous investigations in many other places (Berry 1969; Rees 1970). As such, of course, they might simply reflect the fact that census takers think alike around the world. But the dimensions do indeed appear to index the important interrelationships between regional economic structure, the nature of urban systems, and interdependent processes of urban and regional change (Berry 1969; 1970a). Perhaps, then, they might suggest ways of overcoming the overly simplistic models of spatial structure and process that so far seem to be the afterthoughts of our better theories of developing systems.

## REFERENCES CITED

Berry, Brian J. L. 1969. "Relationships Between Regional Economic Development and the Urban System. The Case of Chile." *Tijd. voor Econ. en Soc. Geog.* 60:283–307.

———. 1970a. "Geography of the United States in the Year 2000?" *Trans. and Papers Inst. Brit. Geog.* 51:21–54.

———. 1970b. "Latent Dimensions of the American Urban System, with International Comparisons." In Berry, B. J. L. ed., *Classification of Cities, New methods and evolving uses.* In press.

Janson, C.-G. 1969. "Some Problems of Ecological Factor Analysis." In Dogan, M., and Rokkan, S., eds., 1969, *Quantitative Ecological Analysis in the Social Sciences.* Cambridge, Mass., The M.I.T. Press, 1969.

Rees, P. 1971. "Factorial Ecology: An Extended Definition, Survey and Critique of the Field." *Econ. Geog. Special issue on Comparative Factorial Ecology.* In press.

| | |
|---|---|
| COPY EDITOR | Philip N. Starbuck |
| DESIGNERS | Carl Kurtz and Rex Wickland |
| PRINTER | The University of Chicago Printing Department |
| PAPER | Natural Sonata Kidskin, 80# text |
| BINDING | Seta 823, arkwright-interlaken, inc. |